# Thinking about Thought

*

*The Structure of Life and the Meaning of Matter*

*

Piero Scaruffi

# Volume 3. Matter

*"Intelligence is not about knowing the answers but about asking the questions"*

*"What we understand is not enough to understand why we understand it"*

Scaruffi, Piero
Thinking about Thought - Matter
All Rights Reserved © 2014 by Piero Scaruffi

ISBN-13: 978-1503362079
ISBN-10: 1503362078

(In the USA only one company is authorized to sell ISBNs: Bowker. And Bowker sells them at an outrageous price when other nations issue ISBNs for free. I consider this as, de facto, one of the most blatant scams in any industry. To protest against this government-sanctioned Bowker ISBN monopoly rip-off, I opted to obtain a free ISBN from Amazon CreateSpace, which will then appear as the publisher of this book, and I encourage all authors and publishers to do the same)

Printed and published in the US

For information: www.scaruffi.com

No part of this book may be reproduced or transmitted in any form or by any means, graphic, electronic or mechanical, including photocopying, recording, taping or by any information storage retrieval system, without the written permission of the author (http://www.scaruffi.com)

# Contents

**SELF-ORGANIZATION AND THE SCIENCE OF EMERGENCE** ............. 5

**THE NEW PHYSICS: THE UBIQUITOUS ASYMMETRY** ......................... 32

**A TIMELINE OF MODERN SCIENCE AND TECHNOLOGY** ................. 137

# Preface

*By the time you finish reading this book you will be a different person. I am not claiming that this book will change the way you think and act. I am simply referring to the fact that the cells in your body, including the neurons of your brain, are continuously changing. By the time you finish reading this book you will "literally" be a different body and a different brain. Every word that you read is having an effect on the connections between your neurons. And every breath you take is pacing the metabolism of your cells. This book is about what just happened to you.*

This volume is one of four in a series titled "Thinking about Thought". See the first volume, "Brain", for the general preface.

## SELF-ORGANIZATION AND THE SCIENCE OF EMERGENCE

### *The Origin of Order*

When Darwin discovered evolution, he also indirectly created the premises for a momentous paradigm shift. Over the centuries, Science had always held that order can only be built by rational agents (e.g., us) who apply a set of fundamental laws of engineering (ultimately, Physics). Scientists such as Galileo and Newton simply refined that model by using more and more sophisticated mathematics. Throughout the theoretical developments of Physics, the fundamental idea remained that, in nature, order needs to be somehow created by external forces. Darwin, instead, showed that order can build itself spontaneously, without any help from a rational outside agent. Evolution is such a process: it is capable of building higher and higher degrees of order starting from almost nothing. As far as Darwin was concerned, this paradigm only applied to Biology, but his idea of spontaneous "emergence" of order can be applied to any natural phenomenon.

More than a century later, Darwin caused a dramatic change in the idea of Physics itself: are splitting the atom and observing distant galaxies the right ways to explain the universe? or should we focus instead on the evolutionary process that gradually built the universe the way it is now? Should we study how things are modified when a force is applied (the vast majority of what Physics does today) or should we deal with how things modify themselves spontaneously? Can Physics ever explain how a tree grows, or how a cloud moves, by bombarding particles with radiation?

The macroscopic phenomena that we observe are more likely to be explained by laws about systems than by laws about particles. The subject of Physics is still the origin of order, but the Darwinian perspective provided a new approach.

Furthermore, order is directly related to information, and Darwin's theory has to do with the creation of information (a new species is a new pattern of information). From this new perspective, Physics may be as much a study of information as it is a study of gravitation or electricity. And the creation of order is inevitably related to the destruction of entropy (or the creation of negative entropy). In the Darwinian view of things, entropy is therefore elevated to a higher rank among physical quantities.

Darwin's laws (unlike the laws of nature claimed by physical sciences) cannot be written down in the form of differential equations. Darwin's laws can only be stated in a "narrative" manner, and any attempt to formalize them resorts to algorithms rather than to equations. Algorithms are fundamentally different from equations in that they are discrete, rather than continuous, occur in steps rather than instantaneously, and can refer to themselves. A Physics based on algorithms would be inherently different from a Physics based on equations.

Finally, Darwin's paradigm is one that is rooted in the concept of organization and that ultimately aims at explaining organization. Indirectly, Darwin helped us understand the elementary fact that the concept of organization is deeply rooted in the physical universe.

Darwin's treatise on the origin of species was indeed a treatise on the origin of order. There lies its monumental importance.

## *Design Without a Designer*

Why do children grow up? Why aren't we born adults? Why do all living things (from organs to ecosystems) have to grow, rather than being born directly in their final configuration?

Darwin's principle was that given a population and fairly elementary rules of how the population can evolve (mainly, variation and natural selection), the population will evolve, and get better and better (adapted) over time. Whether natural selection is really the correct rule is a secondary issue. Darwin's powerful idea was that the target object can be reached not by designing it and then building it, but by taking a primitive object and letting it evolve. The target object will not be built: it will emerge. Trees are not built, they grow. Societies are not built, they form over centuries. Most of the interesting things that we observe in the world are not built, they developed slowly over time. How they happen to be the way they are depends to some extent on the advantages of being the way they are and to some extent on mere chance.

When engineers build a bridge, they don't let chance play with the design and they don't assume that the bridge will grow by itself. They know exactly what the bridge is going to look like and they decide on which day construction will be completed. They know that the bridge is going to work because they can use mathematical formulas. Nature seems to use a different system, in which things use chance to vary, and then variation leads to evolution because of the need for adaptation. By using this system, Nature seems to be able to obtain far bigger and more complex structures than humans can ever dream of building.

It is ironic that, in the process, Nature uses much simpler mathematics. Engineers need to deal with derivatives and cosines. Nature's mathematics (i.e., the mathematics involved in genetic variation) is limited to Arithmetic. Humans have developed a system that is much more complex than anything Nature has ever dreamed of using!

It is stunning that such simple algorithms as used by Nature can produce the complexity of living organisms. Each algorithm can be reduced to even simpler steps. And still the repeated application of those steps eventually yields the complex order of life.

The same theme occurs inside the brain. Neurons exchange simple messages, but the network of those messages over time can produce the very complex behavior of the human mind. That is another simple algorithm that creates complexity.

In both cases the algorithm is simple, but there is a catch. The algorithm is such that every time it ends it somehow remembers the result of its computation and will use it as the starting point for the next run. Species are selected out of the most recently selected species. Neural connections are modified out of the connections already established.

## *Chaos and Complexity*

Both biological and physical sciences need a mathematical model of phenomena of emergence (spontaneous creation of order), and in particular adaptation, as well as a physical justification of their dynamics (which seems to violate physical laws such as the second law of Thermodynamics).

The French physicist Sadi Carnot, one of the founding fathers of Thermodynamics, realized that the statistical behavior of a complex system could be predicted if its parts were all identical and their interactions weak. It was not feasible for Physics to predict the behavior of each single part, but it was relatively easy to predict the statistical behavior of the parts.

At the end of the 19$^{th}$ century, another French physicist, Henri Poincare`, realizing that the behavior of a complex system can become unpredictable if it consists of a few parts that interact strongly, de facto started "chaos" theory. A chaotic system is one in which the effects are not proportional to the causes: in classical (linear) systems a small cause has a small effect, and viceversa; whereas in chaotic (nonlinear) systems there is no proportionality and a small cause can have a big effect, even an effect that spins out of control.

When Poincare showed that the three-body problem does not have a solution because it is a nonlinear system ("On the Three-body Problem and the Equations of Dynamics", 1890), he started a line of thought that over time increasingly found nonlinearity in the universe. Whereas Galileo and Newton had created the impression that the universe was largely a linear system and nonlinearity was a rare exception to the rule, post-Poincare science realized that the reality might just be the opposite.

Formally speaking, a system is said to exhibit the property of chaos if a slight change in the initial conditions results in large-scale differences in the result. This "chaos" opens up a wealth of possibilities for the future of the system. (Note that "chaos" does not mean "disorder": a disordered state cannot be described, whereas a chaotic state can be described by deterministic laws).

The Austrian physicist Heinz von Foerster ("On self-organizing systems and their environments", 1960) noted that self-organization is facilitated by what we call "noise" (or, better, random perturbations). Noise helps a system explore alternative states that were not in its repertory, thereby increasing the chances that the system will find a state of equilibrium. Generalizing: noise causes a system to undergo a state transition that might result in self-organization at a higher level of complexity. Determinism and randomness can be viewed as complementary and symbiotic rather than antagonistic.

The most interesting feature of chaotic systems is that under some circumstances chaotic systems spontaneously "crystallize" into a higher degree of order, i.e. properties begin to emerge. A system must be "complex" enough for any property to "emerge" out of it.

Complexity can be formally defined as nonlinearity. Nonlinearity is best visualized as a game in which playing the game changes the rules. In a linear system an action has an effect that is somewhat proportionate to the effect. In a nonlinear system an action has an effect that causes a change in the action itself. As a matter of fact, the world is mostly nonlinear. The "linear" world studied by classical Physics (Newton's equations are linear) is really just an exception to the rule.

The science of nonlinear dynamics is also known as "chaos theory", from the definition introduced by US physicist James Yorke ("Period Three Implies Chaos", 1975), because unpredictable solutions emerge from nonlinear equations; in other words, they appear to behave in random fashion.

The US mathematician Edward Lorenz ("Deterministic Nonperiodic Flow", 1963) was the first one to focus on the equations about deterministic laws that generate unpredictable behavior. These equations are unpredictable because a slight change in the initial conditions has devastating consequences. In linear equations the effects are proportional to the causes. Linear equations are modular in the sense that one can reduce them to simpler equations and then reassemble the solution. Unfortunately, linearity is an ideal state that usually happens only when a system is close to equilibrium. Real-world systems are rarely close to equilibrium. Anything that is alive, in particular, and anything that is evolving is in a state of non-equilibrium. Nonlinear equations cannot be reduced to simpler equations: the whole is literally more than its parts. This "synergistic" aspect is a sort of internal loop, and is responsible for the property that a small change in the initial conditions can grow exponentially quickly, the so called "butterfly effect" named after an Edward Lorenz lecture ("Does The Flap Of A Butterfly's Wings In Brazil Set Off A Tornado In Texas?", 1972).

Nonlinearity tends to show up when one tries to explain how global behavior emerges from local behavior. For example, it's easy to muster the rules of dynamics and electromagnetism when applied to isolated systems but difficult to predict the weather of a region. It's easy to muster the arithmetic and statistical formulas of economic principles but not to predict what will happen to a nation's economy. The reason is that these complex systems have components that influence each other.

Chaos is not intractable though. As Stephen Smale showed ("Differentiable dynamical systems", 1967), an irregularity can persist in a chaotic system to the point that the system is "stable".

A useful abstraction to describe the evolution of a system in time is that of a "phase space". Our ordinary space has only three dimensions (width, height, depth) but in theory we can think of spaces with any number of dimensions. A phase space has six dimensions, three of which are the usual spatial dimensions while the other three are the components of velocity along those spatial dimensions. In ordinary 3-dimensional space, a "point" can only represent the position of a system. In 6-dimensional phase space, a point represents both the position and the motion of the system. The evolution of a system is represented by some sort of shape in phase space.

An "attractor" is a region of phase space where the system is doomed to end up eventually. For a linear system the attractor basically describes its typical behavior. A point attractor is the state in which the system ends its motion. A periodic attractor is the state that a system returns to periodically. Chaotic systems do not have point or periodic attractors, but chaotic systems may exhibit attractors that are shapes with fractional dimension. For example, the dimensions of attractors of planetary orbits are integers, whereas the attractors of asteroidal orbits have fractional dimensions.

If the region that these "strange attractors" occupy is finite, then, by definition, the behavior of the chaotic system is not truly random. The first "strange attractor" was "discovered" by the US meteorologist Edward Lorenz ("Deterministic Nonperiodic Flow", 1963): it was due to a nonlinear system that evolves over time in a non-repeating pattern. In phase space, its evolution looks like an infinite series of ever-changing patterns that seem to be attracted to a point but never repeat the same trajectory around it.

The US physicist Robert Shaw ("Strange Attractors, Chaotic Behavior, and Information Flow", 1981) realized that strange attractors increase entropy, i.e. "create" information. They create information about the behavior of a chaotic system where no information about it was available. Information is continuously being created by "chaotic" phenomena, which are pervasive in nature.

Nonlinear processes are ubiquitous. They are processes of emergent order and complexity, of how structure arises from the interaction of many independent units. These processes recur at every level, from morphology to behavior. At every level of science (including the brain and life) the spontaneous emergence of order, or self-organization of complex systems, is a common theme.

One feature of chaos, discovered by the Australian mathematician Robert May (""Simple Mathematics Models with very complicated dynamics", 1976), is "self-similarity", the fact that the "chaotic" behavior sometimes embeds an exact replica of itself. By studying self-similarity, the Polish-born mathematician Benoit Mandelbrot came up with the idea of fractal geometry ("Fractal Objects", 1975). Euclidean geometry fails to capture the essence of ordinary natural shapes that are way more complex than straight lines or perfect circles. Mandelbrot introduced fractional dimensions to express the degree of "irregularity" of a shape; for example, of a coastline. Then the scale at which one carries out a measurement becomes important: Mandelbrot found that Portuguese and Spanish books came up with different lengths of the Spanish-Portuguese border. When a line is irregular, there is no number that is absolutely correct: it depends on how small the "ruler" is that measures it. If one uses a 1km-long ruler, one obtains a value for the length of the border. If one uses a one-meter long ruler, a much bigger value is obtained. Euclidean geometry does not depend on scale: the properties of a triangle or of a circle do not depend on the scale at which they are observed. Fractal geometry depends on the scale.

The Italian physicist Luciano Pietronero even argued that the universe is fundamentally fractal ("The fractal structure of the universe", 1987).

The US physicist Mitchell Feigenbaum discovered a new universal constant that applies to chaotic systems by analyzing how a nonlinear system becomes chaotic ("Quantitative Universality for a Class of Nonlinear Transformations", 1978). At some point the behavior of the system splits into two (a "bifurcation") and then into four and so forth, at an increasingly faster rate. It is this acceleration of bifurcations that defines "chaos". Feigenbaum discovered that the ratio between the bifurcations is 4.669. There is order even in chaos.

Darwin's vision of natural selection as a creator of order is probably not sufficient to explain all the spontaneous order exhibited by both living and inanimate matter. There might be other physical principles at work.

Arthur Koestler and Stanley Salthe showed how complexity entails hierarchical organization. Von Bertalanffy's general systems theory, Haken's synergetics, and Prigogine's non-equilibrium Thermodynamics belong to the class of theories that extend Physics to dynamic systems.

These theories have in common the fact that they deal with self-organization (how collections of parts can produce structures) and try to provide a unified view of the universe at different levels of organization (from living organisms to physical systems to societies).

The drawback in any study of complex systems is that there is no commonly accepted definition of complexity and method of measuring complexity. Andrey Kolmogorov's complexity is not a biologist's complexity. A biologist does not know how to compare the complexity of a tomato with the complexity of the stock market.

## *Holarchies*

The Hungarian writer Arthur Koestler brought together a wealth of biological, physical, anthropological and philosophical notions to construct a unified theory of open hierarchical systems that goes beyond the reductionist method favored by Science since Descartes.

Language has to do with a hierarchical process of spelling out implicit ideas in explicit terms by means of rules and feedback. Organisms and societies also exhibit the same hierarchical structure. In these hierarchies, each intermediary entity ("holon") functions as a self-contained whole relative to its subordinates and as one of the dependent parts of its superordinates. Each holon tends to persist and assert its pattern of activity.

Koestler thought that, wherever there is life, it must be hierarchically organized. He argued that life exhibits an integrative property (that manifests itself as symbiosis) that enables the gradual construction of complex hierarchies out of simple holons. In nature there are no separated, indivisible, self-contained units. An "individual" is an oxymoron. An organism, instead, is a hierarchy of self-regulating holons (a "holarchy") that work in conjunction with their environment. Holons at the higher levels of the hierarchy enjoy progressively more degrees of freedom and holons at the lower levels of the hierarchy have progressively less degrees of freedom. Moving up the hierarchy, we encounter more and more complex, flexible and creative patterns of activity. Moving down the hierarchy behavior becomes more and more mechanized.

Hierarchical processes of the same nature can be found in the development of the embryo, in the evolution of species and in consciousness itself. The latter should be analyzed not in the context of the mind/body dichotomy but in the context of a multi-leveled hierarchy and of degrees of consciousness.

They all share common themes: a tendency towards integration (a force that is inherent in the concept of hierarchic order, even if it seems to challenge the second law of Thermodynamics as it increases order), an openness at the top of the hierarchy (towards higher and higher levels of complexity) and the possibility of infinite regression.

## *Hierarchies from Complexity*

The US zoologist Stanley Salthe developed a multi-dimensional theory: an ontology of the world, a hierarchical representation of the world and a model of the evolution of the world.

Salthe inherits a definition of complexity from the US biologist Howard Pattee: complexity is the result of interactions between physical and symbolic systems. A physical system is dependent on the rates at which processes occur, whereas a symbolic system is not. Symbolic systems frequently serve as constraints applied to the operation of physical systems, and frequently appear as products of the activity of physical systems (e.g., the genome in a cell). A physical system can be said to be "complex" when a part of it functions as a symbolic system (as a representation, and therefore as an observer) for another part of it.

John Von Neumann was possibly the first scientist to realize that it takes a certain degree of complexity for physical molecules to enter the cycle of Darwinian open-ended evolution. This degree of complexity is determined by description-based reproduction. Once there is description-based reproduction, then there is Darwinian evolution. Von Neumann thus made a distinction between "quiescent" symbolic description and "active" physical dynamics, between descriptions and constructions, between, ultimately, semiotic processes (information and codes) and physical systems (energy-matter and forces). Pattee believed that this distinction occurs at all levels of organization in the universe, not only at the level of genes and proteins. At the highest level this distinction translates into the distinction between the physical systems that populate the universe and the semiotic process of measurement (that codes a dynamical state into quiescent symbols). The genotype-phenotype distinction an instance of the origin of symbol systems from physical systems. What causes normal physical molecules to start functioning as descriptions? Pattee called this phenomenon, that separates description from construction and that occurs at all levels of organization, the "epistemic cut". Its origin lies not in the chemical properties of cells but in the complex relationships established within a complex hierarchical system.

Salthe was also influenced by the metaphysics of the US philosopher Justus Buchler. His "natural complex" is pretty much anything that one can think of, whether organism, concept or conscious event. On the other hand, an "order" is a multiplicity that becomes a unity by virtue of its internal organization (by virtue of the pattern of relatedness among its components). Buchler's "principle of ordinality" states that every natural complex is an order. Basically, the principle of ordinality asserts that every complex must be constituted by other complexes, and that every complex must be one of the constituents of some other complex. Every complex is relative to some other complex, is conditioned by and conditions other complexes.

The Argentine philosopher Mario Bunge saw the universe not as a heap of things but as a system composed of interconnected systems of various kinds (physical, biological, economic, political, cultural). Bunge's systemism offered an alternative to both individualism and holism, allowing for both individual identity and collective organization. Bunge argued that a system is defined by "composition" (what it is made of), "environment" (what surrounds it), "structure" (what holds it together) and "mechanism" (how it operates).

In Salthe's theory, the world is viewed as a determinate machine of unlimited complexity. Within complexity, discontinuities arise. The basic structure of this world must allow for complexity that is spontaneously stable and that can be broken down to things divided by boundaries. The most natural way for the world to satisfy this requirement is to employ a hierarchical structure, which is also implied by Buchler's principle of ordinality: Nature is a hierarchy of entities existing at different levels of organization. Hierarchical structure turns out to be a consequence of complexity.

Entities of the hierarchy are defined by four attributes: boundaries, scale, integration, continuity. An entity has size, is limited by boundaries, and consists of an integrated system, which varies continuously in time.

Entities at different levels interact through mutual constraints, each constraint carrying information for the level it operates upon. A process can be described by a triad of contiguous levels: the one it occurs at, its context (what Bunge called "environment") and its causes (Bunge's "structure"). In general, a lower level provides initiating conditions for a process and an upper level provides boundary conditions. Representing a dynamic system hierarchically requires a triadic structure.

Aggregation occurs upon differentiation. Differentiation interpolates levels between the original two and the new entities aggregate in such a way that affects the structure of the upper levels: every time a new level emerges, the entire hierarchy must reorganize itself.

These abstract principles also apply to biological evolution. Over time, Nature generates entities of gradually more limited scope and more precise form and behavior. This process populates the hierarchy of intermediate levels of organization as the hierarchy spontaneously reorganizes itself. The same model applies to all open systems, whether organisms or ecosystems or planets.

Basically, Salthe aims at reformulating Biology on development rather than on evolution. His approach is non-Darwinian to the extent that development, and not evolution, is assumed to be the fundamental process in self-organization. Evolution, in his opinion, is merely the result of a margin of error.

Salthe's grand theory of nature turns out to be essentially a theory of change, which turns out to be essentially a theory of emergence.

### *General Systems Theory*

"General Systems Theory" was born before Cybernetics, and cybernetic systems are merely a special case of self-organizing systems; but General Systems Theory took longer to establish itself. It was conceived in the 1930s by the Austrian biologist Ludwig Von Bertalanffy. His ambition was to create a "universal science of organization". His legacy is to have started "system thinking", thinking about systems as systems and not as mere aggregates of parts.

The classical approach to the scientific description of a system's behavior (whether in Physics or in Economics) can be summarized as the search for "isolatable causal trains" and the reduction to atomic units. This approach is feasible under two conditions: 1. Thatt the interaction among the parts of the system be negligible and 2. That the behavior of the parts be linear. Von Bertalanffy's "systems", on the other hand, are those entities ("organized complexities") that consist of strongly interacting parts, usually described by a set of nonlinear differential equations. Systems Theory studies principles that apply to all systems, i.e. properties that apply to any entity qua system.

Basic concepts of Systems Theory are, for example, the following: every whole is based upon the competition among its parts; individuality is the result of a never-ending process of progressive centralization whereby certain parts gain a dominant role over the others.

General Systems Theory mainly studies "wholes", which are characterized by such holistic properties as hierarchy, stability, teleology.

General Systems Theory looks for laws that can be applied to a variety of fields (i.e., for an isomorphism of laws in different fields), particularly in the biological, social and economic sciences (but even to history and politics).

"Open Systems Theory" is a subset of General Systems Theory. Because of the second law of Thermodynamics, a change in entropy in closed systems is always positive: order is continually destroyed. On the other hand in open systems (such as living systems) entropy production due to irreversible processes is balanced by import of negative entropy (as in all living organisms).

A living organism can be viewed as a hierarchical order of open systems, where each level maintains its structure thanks to continuous change of components at the next lower level. Living organisms maintain themselves in spite of continuous irreversible processes and even proceed towards higher and higher degrees of order.

The "theory of natural systems" of the Hungarian philosopher Ervin Laszlo is a theory of the invariants of organized complexity. It is centered on the concept of "ordered whole", whose structure is defined by a set of constraints. Laszlo adopts a variant of the principle of self-organization formulated by the British neurologist Ross Ashby ("Principles of the self-organizing dynamic system", 1947), according to which any isolated natural system subject to constant forces is inevitably inhabited by "organisms" that tend towards stationary or quasi-stationary non-equilibrium states. Natural systems sharing an environment tend to organize in hierarchies. The set of such systems tends to become itself a system, its subsystems providing the constraints for the new system.

In Laszlo's view, the combination of internal constraints and external forces yields adaptive self-organization. Natural systems evolve towards increasingly adapted states, corresponding to increasing complexity (and negative entropy).

Order emerges at the atomic ("micro-cybernetics"), organismic ("bio-cybernetics") and social levels ("socio-cybernetics").

The system-oriented approach can also address a particular class of natural systems: cognitive systems. The mind, just like any other natural system, exhibits a holistic character, adaptive self-organization, and hierarchies, and can be studied with the same tools used for all other natural systems ("psycho-cybernetics").

Laszlo views the dynamics of the universe as driven by "third-state systems". First-state systems are systems in equilibrium. Second-state systems are systems in near equilibrium. Third-state systems are non-linear systems that are farthest from equilibrium. Third-state system must import energy in order to survive, and, in doing so, they end up creating new order, at higher and higher levels of complexity. These systems tend to form hyper-cycles, and Laszlo calls this tendency "convergence". It is convergence that led to the formation of galaxies, to the evolution of more complex forms of life, to the birth of consciousness. Laszlo's convergence seems to act like a universal force that endlessly destroys order and rebuilds it at a higher level.

### *Synergetics*

"Synergetics", as developed by the German physicist Hermann Haken, is a theory of pattern formation in complex systems. It tries to explain structures that develop spontaneously in nature.

Synergetics studies cooperative processes of the parts of a system far from equilibrium that lead to an ordered structure and behavior for the system.

Haken's favorite example was the laser: how do the atoms of the laser agree to produce a single coherent wave flow? The answer is that the laser is a self-organizing system far from equilibrium (what Prigogine would call a dissipative structure).

A "synergetic" process in a physical system is one in which, when energy is pumped into the system, some macroscopic structure emerges from the disorderly behavior of the large number of microscopic particles that make up the physical system. As energy is pumped into the system, initially nothing seems to happen, other than additional excitation of the particles, but then the system reaches a threshold beyond which structure suddenly emerges. The laser is such a synergetic process: a beam of coherent light is created out of the chaotic movement of particles. What happens is that energy pushes the system of particles beyond a threshold, and suddenly the particles start behaving harmoniously..

Synergetics revolves around a number of technical concepts: compression of the degrees of freedom of a complex system into dynamic patterns that can be expressed as a collective variable; behavioral attractors of changing stabilities; and the appearance of new forms as non-equilibrium phase transitions.

Systems at instability points (at the "threshold") are driven by a "slaving principle": long-lasting quantities (the macroscopic pattern) can enslave short-lasting quantities (the chaotic particles), and they can force order on them (thereby becoming "order parameters").

The system exhibits both a stable "mode", which is the chaotic motion of its particles, and an unstable "mode", which is the macroscopic structure and behavior of the whole system. Close to instability, stable modes are "enslaved" by unstable modes and can be ignored. Instead of having to deal with millions of chaotic particles, one can focus on the macroscopic quantities. De facto, the degrees of freedom of the system are reduced.

The dynamic equations for such a system reflect the interplay between stochastic forces ("chance") and deterministic forces ("necessity").

Synergetics applies to systems driven far from equilibrium, where the classic concepts of Thermodynamics are no longer adequate. It expresses the fact that order can arise from chaos and can be maintained by flows of energy/matter.

### *Hypercycles*

The German chemist Manfred Eigen was awarded the Nobel Prize in 1967 for discovering that very short pulses of energy could trigger extremely fast chemical reactions. In the following years, he started looking for how very fast reactions could be used to create and sustain life.

Indirectly, he ended up studying the behavior of biochemical systems far from equilibrium.

Eventually, Eigen came up with the concept of an "hypercycle". A hypercycle is a cyclic reaction network, i.e. a cycle of cycles of cycles (of chemical reactions). Then he argued that life can be viewed as the product of a hierarchy of such hypercycles.

A catalyst is a substance that favors a chemical reaction. When enough energy is provided, some catalytic reactions tend to combine to form networks, and such networks may contain closed loops, called catalytic cycles.

If even more energy is pumped in, the system moves even farther from equilibrium, and then catalytic cycles tend to combine to form closed loops of a higher level, or hypercycles, in which the enzymes produced by a cycle act as catalysts for the next cycle in the loop. Each link of the loop is now a catalytic cycle itself.

Eigen showed that hypercycles are capable of self-replication, which may therefore have been a property of nature even before the invention of living organisms.

Hypercycles are capable of evolution through more and more complex stages. Hypercycles compete for natural resources and are therefore subject to natural selection.

The hypercycle falls short of being a living system because it defines no "boundary": the boundary is the container where the chemical reaction is occurring. A living system, on the other hand, has a boundary that is part of the living system (e.g., the skin).

Catalysis is the phenomenon by which a chemical reaction is sped up: without catalysis, all processes that give rise to life would take a lot longer, and probably would not be fast enough for life to happen. Then Eigen shows that they can be organized into an autocatalytic cycle, i.e. a cycle that is capable of self-reproducing: this is the fundamental requirement of life. A set of autocatalytic cycles gets, in turn, organized into a catalytic hypercycle. This catalytic hypercycle represents the basic form of life.

Formally: "hypercycles" are a class of nonlinear reaction networks. They can originate spontaneously within the population of a species through natural selection and then evolve to higher complexity by allowing for the coherent evolution of a set of functionally coupled self-replicating entities. A hypercycle is based on nonlinear autocatalysis, which is a chain of reproduction cycles, which are linked by cyclic catalysis, i.e. by another autocatalysis. A hypercycle is a cycle of cycles of cycles.

Eigen's model explains the simultaneous unity (due to the use of a universal genetic code) and diversity (due to the "trial and error" approach of natural selection) in evolution. This dual process started even before life was created. Evolution of species was preceded by an analogous stepwise process of molecular evolution.

Evolution itself turns out to be inevitable: given a set of self-reproducing entities that feed on a common and limited source of energetic/material supply, evolution will spontaneously appear. Evolution is a direct consequence of the dynamics of self-reproducing systems.

That said, not all systems are suitable for becoming successful biological systems. Systems can be classified in four groups according to their stability with respect to fluctuations: stable systems (the fluctuations are self-regulating), indifferent systems (the fluctuations have no effect), unstable systems (self-amplification of the fluctuations) and variable systems (which can be in any of the previous states). Only the last type is suitable for generation of biological information because it can play all the best tactics: indifference towards a broad mutant spectrum, stability towards selective advantages and instability towards unfavorable configurations. In other words, it can take the most efficient stance in the face of both favorable and adverse situations.

## *Autonomous Systems*

The Chilean neurologist Francisco Varela adapted Humberto Maturana's biological ideas to his theory of autonomous systems. He merged the themes of autonomy of natural systems (i.e. internal regulation) and their informational abilities (i.e., cognition) into the theme of a system maintaining an identity and interacting with the rest of the world.

The organization of a system is the set of relations that define it as a unity. The structure of a system, on the other hand, is the set of relations among its components. Components and relations among them may change over time without necessarily changing the overall organization. For example, a machine can be implemented by different sets of components and relations among them.

"Homeostatic" systems are systems that keep the values of their variables within a small range of values.

An "autopoietic" system is a homeostatic system that continuously generates its own organization, by continuously producing components that are capable of reproducing the organization that created them.

Autopoietic systems turn out to be autonomous, to have an identity, to be unities, and to compensate external perturbations with internal structural changes.

Living systems are autopoietic systems in the physical space. Self-reproduction can only occur in autopoietic systems, and evolution is a direct consequence of self-reproduction.

By definition, an autonomous system is organizationally closed. The cognitive domain of an autonomous system is the domain of interaction that it can enter without losing that closure.

An autonomous system always exhibits two aspects: it specifies the distinction between self and non-self, and deals with its environment in a cognitive fashion. Therefore, every autonomous system (ecosystems, societies, brains, even conversations) is a "mind".

## *A Science of Prisms*

In the 1970s the US inventor Buckminster Fuller developed a theory, also called "Synergetics", that approached systems from a holistic perspective that is basically the opposite of the reductionist perspective of Physics. Fuller's philosophy was inspired by one of his own inventions, the "geodesic" dome (1954), a structure that exploits a very efficient way of enclosing space and that gets stronger as it gets larger.

"Synergy" is the behavior of a whole that cannot be explained by the parts taken separately. For example, a star attracts a planet: humans could not have predicted this by simply studying the two bodies separately. Synergetics, therefore, studies a system in a holistic (rather than reductionist) fashion. The way it does this is by focusing on form rather than internal structure. Because of its emphasis on shape, Synergetics is, de facto, a branch of Geometrics, the discipline of configurations (or patterns).

Synergetics employs 60-degree coordination instead of the usual 90-degree coordination. The triangle (and the tetrahedron) instead of the square (and the cube) is the fundamental geometric unit. The tetrahedron is the minimal system with the fewest possible points.

Fuller argued that reality is not made of "things", but of angle and frequency events. All experience can be reduced to only angles and frequencies. Fuller found "prisms" to be ubiquitous in nature and in culture. All systems contained in the universe are polyhedral, "universe" being the collection of all experiences of all individuals.

Synergetics rediscovers, in an almost mystical way, most of traditional science, but mainly through topological considerations (with traditional topology extended to "omnitopology"). For example, Synergetics proves that the universe is finite and expanding, and that Planck's constant is a "cosmic relationship". Synergetics unifies Physics and Metaphysics.

## *Dissipative Systems*

In 1951 Boris Belousov discovered what came to be known as the Belousov–Zhabotinsky (BZ) reaction, a class of chemical reactions that spontaneously generate highly-ordered structures far from equilibrium, behaving like a sort of spiral. Similar spiral-like behavior would be later discovered in many other natural phenomena, for example by the US astronomers Wendy Freedman and Barry Madore ("Self-Organizing Structures." 1987).

The Belgian (but Russian-born) physicist Ilya Prigogine showed that all biological systems actually belong to the same class of systems: they are all dissipative systems.

Classical Physics describes the world as a static and reversible system that undergoes no evolution, whose information is constant in time. Classical Physics is the science of being. Thermodynamics, instead, describes an evolving world in which irreversible processes occur. Thermodynamics is the science of becoming.

The second law of Thermodynamics, in particular, describes the world as evolving from order to disorder, while biological evolution is about the complex emerging from the simple (i.e. order arising from disorder). While apparently contradictory, these two views show that irreversible processes are an essential part of the universe.

Furthermore, conditions far from equilibrium foster phenomena such as life that classical Physics does not cover at all.

Irreversible processes and non-equilibrium states turn out to be fundamental features of the real world.

Prigogine distinguishes between "conservative" systems (which are governed by the three conservation laws for energy, translational momentum and angular momentum, and which give rise to reversible processes) and "dissipative" systems (subject to flows of energy and/or matter). The latter give rise to irreversible processes.

The theme of science is order. Order can come either from equilibrium systems or from non-equilibrium systems that are sustained by a constant source (or, equivalently, by a persistent dissipation) of matter/energy. In the latter systems, order is generated by the flow of matter/energy. All living organisms (including the biosphere as a whole) are non-equilibrium systems.

Prigogine proved that, under special circumstances, the distance from equilibrium and the nonlinearity of a system drive the system to ordered configurations (of the kind that Belousov had discovered), i.e. create order. The science of being and the science of becoming describe dual aspects of Nature.

What is needed is a combination of factors that are exactly the ones found in living matter: a system made of a large collection of independent units which are interacting with each other; a flow of energy through the system that drives the system away from equilibrium; and nonlinearity. Nonlinearity expresses the fact that a perturbation of the system may reverberate and have disproportionate effects.

Non-equilibrium and nonlinearity favor the spontaneous development of self-organizing systems, which maintain their internal organization, regardless of the general increase in entropy, by expelling matter and energy in the environment.

When such a system is driven away from equilibrium, local fluctuations appear. This means that the system gets very unstable in some places. Localized tendencies to deviate from equilibrium are amplified. When a threshold of instability is reached, one of these runaway fluctuations is so amplified that it takes over as a macroscopic pattern. Order appears from disorder through what are initially small fluctuations within the system. Most fluctuations die along the way, but some survive the instability and carry the system beyond the threshold: those fluctuations "create" new form for the system. Fluctuations become sources of innovation and diversification.

The potentialities of nonlinearity are dormant at equilibrium but are revealed by non-equilibrium: multiple solutions appear and therefore diversification of behavior becomes possible.

Technically speaking, nonlinear systems driven away from equilibrium can generate instabilities that lead to "bifurcations" (and "symmetry breaking" beyond bifurcation). When the system reaches the bifurcation point, it is impossible to determine which path it will take next. Chance rules. Once the path is chosen, determinism resumes.

The multiplicity of solutions in nonlinear systems can even be interpreted as a process of gradual "emancipation" from the environment.

Most of Nature is made of such "dissipative" systems, of systems subject to fluxes of energy and/or matter. Dissipative systems conserve their identity thanks to the interaction with the external world. In dissipative structures, non-equilibrium becomes a source of order.

In general, self-organization is the spontaneous emergence of ordered structure and behavior in open systems that are in a state far from equilibrium described mathematically by nonlinear equations.

These considerations apply to living organisms, which are prime examples of dissipative structures in non-equilibrium. Prigogine's theory explains how life can exist and evolution can work towards higher and higher forms of life. A "minimum entropy principle" characterizes living organisms: stable near-equilibrium dissipative systems minimize their rate of entropy production.

Catastrophe and chaos theories are special cases of nonlinear non-equilibrium systems.

*Catastrophe Theory*

Catastrophe theory, originally formulated in the 1960s by the French mathematician René Thom ("A dynamic theory of morphogenesis", 1966) and popularized ten years later by the work of the British mathematician Erich Zeeman, became a widely used tool for classifying the solutions of nonlinear systems in the neighborhood of stability breakdown.

In the beginning, Thom was interested in structural stability in topology (stability of topological form) and was convinced of the possibility of finding general laws of form evolution regardless of the underlying substance of form, as already stated at the beginning of the century by the British biologist D'Arcy Thompson.

Thom's goal was to explain the "succession of form". Our universe presents us with forms (that we can perceive and name). A form is defined, first and foremost, by its stability: a form lasts in space and time. Forms change. The history of the universe, insofar as we are concerned, is a ceaseless creation, destruction and transformation of form. Life itself is, ultimately, creation, growth and decaying of form.

Every physical form is represented by a mathematical quantity called an "attractor" in a space of internal variables. If the attractor satisfies the mathematical property of being "structurally stable", then the physical form is the stable form of an object. Changes in form, or morphogenesis, are due to the capture of the attractors of the old form by the attractors of the new form. All morphogenesis is due to the conflict between attractors.

The universe of objects can be divided into domains of different attractors. Such domains are separated by shock waves. Shock wave surfaces are singularities called "catastrophes". A catastrophe is a state beyond which the system is destroyed in an irreversible manner. Technically speaking, the "ensembles de catastrophes" are hypersurfaces that divide the parameter space in regions of completely different dynamics.

The bottom line is that dynamics and form become dual properties of nonlinear systems.

Thom proves that in a 4-dimensional space there exist seven types of elementary catastrophes. Elementary catastrophes include: "fold", destruction of an attractor, which is captured by a lesser potential; "cusp", bifurcation of an attractor into two attractors; etc. From these singularities, more and more complex catastrophes unfold, until the final catastrophe. Elementary catastrophes are "local accidents". The form of an object is due to the accumulation of many of these "accidents".

What catastrophe theory does is to "geometrize" the concept of "conflict". This theory is a purely geometric theory of morphogenesis, Its laws are independent of the substance, structure and internal forces of the system.

### *The Origin of Regularity*

Prigogine's bifurcation theory is a descendent of the theory of stability initiated by the Russian mathematician Aleksander Lyapounov. Thom's catastrophe theory is a particular case of bifurcation theory, so they all belong to the same family. They all elaborate on the same theorem, namely Lyapounov's theorem of 1892: for isolated systems, thermodynamic equilibrium is an attractor of nonequilibrium states.

Then the story unfolds, leading to dissipative systems and eventually to the reversing of Thermodynamics' fundamental assumption, the destruction of structure. Order emerges from the very premises that seem to deny it.

### *Simplexity and complicity*

The British biologist Jack Cohen and the British mathematician Ian Stewart studied how the regularities of nature (from Cosmology to Quantum Theory, from Biology to Cognitive Psychology) emerge from the underlying chaos and complexity of nature: "emergent simplicities collapse chaos". They argued that external constraints are fundamental in shaping biological systems (DNA does not uniquely determine an organism) and defined new concepts: "simplexity" (the tendency of simple rules to emerge from underlying disorder and complexity) and "complicity" (the tendency of interacting systems to co-evolve leading to a growth of complexity). Simplexity is a "weak" form of emergence, and is ubiquitous. Complicity is a stronger form of emergence, and is responsible for consciousness and evolution. "Simplexity merely explores a fixed space of the possible... complicity enlarges it."

Emergence is the rule, not the exception, and it is shaped by simplexity and complicity.

### *The Edge of Chaos*

The US computer scientist Chris Langton (who organized the first Artificial Life conference in 1987) showed that physical systems achieve the prerequisites for the emergence of computation (i.e., transmission, storage, modification) in the vicinity of a phase transition, "at the edge of chaos" ("Computation at the Edge of Chaos", 1990). When a system is in a highly ordered state, the transfer and modification of information becomes impossible. When a system in in a highly chaotic state, no pattern of information can persist over time. As systems become less orderer and less chaotic, a compromise is reached, whereby information can exist, can be manipulated, can be transferred. In technical terms, information becomes an important factor in the dynamics of cellular automata in the vicinity of the phase transition between periodic and chaotic behavior.

The general idea is that some systems undergo transformations, and while they transform they constantly move from order to chaos and back. This transition is similar to the "phase transitions" undergone by a substance when it turns liquid or solid or fluid. When ice turns into water, the atoms have not changed, but the system as a whole has undergone a phase transition. Microscopically, this means that atoms are behaving in a different way. The transition of a system from chaos to order and back is similar in that the system is still made of the same parts, but they behave in a different way.

The state between order and chaos (the "edge of chaos") is sometimes a very "informative" state, because the parts are not as rigidly assembled as in the case of order and, at the same time, they are not as loose as in the case of chaos. The system is stable enough to keep information and unstable enough to dissipate it. The system at the edge of chaos is both a storage and a broadcaster of information.

At the edge of chaos, information can propagate over long distances without decaying appreciably, thereby allowing for long-range correlation in behavior: ordered configurations do not allow for information to propagate at all, and disordered configurations cause information to quickly decay into random noise.

A fundamental connection therefore exists between computation and phase transition.

The edge of chaos is where the system can perform computation, can metabolize, can adapt, can evolve. In a word: these systems can be alive.

Basically, Langton proved that Physics can support life only in a very narrow boundary between chaos and order. In that locus it is possible to build "organisms" that will settle into recurring patterns conducive to an orderly transmission of information.

Langton's theory related phase transitions, computation and life: he built a bridge to link together Thermodynamics, Information Theory and Biology.

Likewise, the US physicist Murray Gell-Mann argued that living organisms dwell at the edge of chaos, as they exhibit order and chaos at the same time, and they must exhibit both in order to survive. Living organisms are complex adaptive systems that retrieve information from the world, find regularities, compress them into a schema to represent the world, predict the evolution of the world and prescribe behavior for themselves. The schema may undergo variants that compete with one another. Their competition is regulated by feedback from the real world under the form of selection pressure. Disorder is useful for the development of new behavior patterns that enable the organism to cope with a changing environment.

### *Complex Systems*

The US biologist Stuart Kauffman views the dynamics of "complex systems" as a manifestation of the fundamental force that counteracts the universal drift towards disorder required by the second law of Thermodynamics.

His idea is that Darwin was only half right: systems do evolve under the pressure of natural selection, but their quest for order is helped by a property of our universe, the property that "complex" systems just tend to organize themselves. Darwin's story is about the power of chance: by chance life developed and then evolved. Kauffman's story is about destiny: life is the almost inevitable result of a process inherent in nature.

Kauffman's starting point was that cells behave like mathematical networks.

In the early 1960s the French biologist Jacques Monod and others discovered gene regulation: genes are assembled not in a long string of instructions but in "genetic circuits". Within the cell, there are regulatory genes whose job is to turn on or off other genes. Therefore genes are not simply instructions to be carried out one after the other. Genes realize a complex network of messages. A regulatory gene may trigger another regulatory gene that may trigger another gene...etc. Each gene is typically controlled by two to ten other genes. Turning on just one gene may trigger an avalanche of effects.

The genetic program is not a sequence of instructions but rather a regulatory network that behaves like a self-organizing system.

By using a computer simulation of a cell-like network, Kauffman proved that, in any organism, the number of cell types must be approximately the square root of the number of genes.

He basically started where Langton ended. His "candidate principle" states that organisms change their interactions in such a way to reach the boundary between order and chaos.

For example, the Danish physicist Per Bak ("Self-organized criticality", 1987) studied a pile of sand, whose collapse under the weight of a new randomly placed grain is unpredictable. However, when it happens, the pile reorganizes itself. While it is impossible to predict if a particular grain will cause an avalanche, the size of these avalanches is distributed according to a "power law". No external force is shaping the pile of sand: it is the pile of sand that organizes itself.

Further examples include any ecosystem (in which organisms live at the border between extinction and overpopulation), the price of a product (which is defined by supply and demand at the border of where nobody wants to buy it and where everybody wants to buy it). Evolution proceeds towards the edge of chaos. Systems on the boundary between order and chaos have the flexibility to adapt rapidly and successfully.

Natural selection and self-organization complement each other: they create complex systems poised at the edge between order and chaos, which are fit to evolve in a complex environment. At all levels of organization, whether of living organisms or ecosystems, the target of selection is a type of adaptive system at the edge between chaos and order.

In the 1930s the US biologist Sewall Wright had introduced the concept of "fitness landscapes" ("The roles of mutation, inbreeding, crossbreeding and selection in evolution", 1932). Fitness is the replication rate of a genotype. A fitness landscape is a distribution of fitness values over the space of genotypes. In other words, the fitness landscape describes all possible genotypes, their degree of similarity and their fitness values. Fitness is related to height in the landscape. Genotypes that are very similar are close to each other in the landscape.

Evolution is the traversing of a fitness landscape. Peaks represent optimal fitness. Populations wander through the landscape, driven by mutation, selection and drift, in their search for peaks. Kauffman showed that the best strategy for reaching the peaks occurs at the phase transition between order and disorder, or, again, at the edge of chaos. The same model applies to other biological phenomena and even nonbiological phenomena, and may therefore represent a universal law of nature.

Adaptive evolution can be represented as a local "hill-climbing search" that converges via fitter mutants toward some local or global optimum. Adaptive evolution occurs on rugged (multipeaked) fitness landscapes.

The very structure of these landscapes implies that "radiation" and "stasis" are inherent features of adaptation. The Cambrian explosion and the Permian extinction (famous paradoxes of the fossil record) may be the natural consequences of inherent properties of rugged landscapes.

Kauffman also noted how complex (nonlinear dynamic) systems which interact with the external world are bound to "classify" and "know" their world through their attractors.

Kauffman's view of life can be summarized as follows: autocatalytic networks (networks that feed themselves) arise spontaneously; natural selection brings them to the edge of chaos; a genetic regulatory mechanism accounts for metabolism and growth; attractors lay the foundations for cognition.

The main theme of Kauffman's research is the ubiquitous trend towards self-organization. This trend causes the appearance of "emergent properties" in complex systems. One such property is life. The requirements for order to emerge are far easier than traditionally assumed.

There is order for free.

Far from equilibrium, systems organize themselves. The way they organize themselves is such that it creates systems at higher levels, which in turn tend to organize themselves. Atoms organize in molecules that organize in autocatalytic sets that organize in living organisms that organize in ecosystems.

The whole universe may be driven by a principle similar to autocatalysis. The universe may be nothing but a hierarchy of autocatalytic sets.

Of course, one possible objection to the whole theory of "self-organizing" systems is that no system truly "self-organizes": they all depend on external energy. Thus one could claim that it is the external energy that organizes them. Self-organizing systems, strictly speaking, do not exist. Only the universe as a whole can be said to be truly self-organizing.

### *Chaos as the Creator*

Anorther corollary to this vision of multi-layered self-organization is about the role played by chaos. Gregory Bateson, Francisco Varela, Chris Langton and Stuart Kauffman assign a creative role to chaos. A living organism is a self-organizing sysetm, a system that tries to maintain order inside. Outside there is a chaotic world, and that is the reason why the self-organization is a continuous, endless process. At that same time it is that chaotic "outside" that fosters creativity, growth and evolution. Too much "chaos" might kill the living organism, but, whenever the organism if capable of responding to chaos, the result is actually an increase in complexity and "intelligence".

### *Synchronized Oscillators*

There is a way to look at processes of self-organization that is alternative to the view based on phase transitions: it is based on synchronized oscillators. Self-organization of a system implies that every part of that system is somewhat "synchronized" in realizing some kind of order.

Clocks and cycles are pervasive in our universe, ranging from the biological clocks of the tiniest organisms to the cycles of human history, from the cycles of electrons around the nucleus to the cycles of planets around their stars. The order of these cycles emerges from the synchronized behavior of "coupled oscillators". The tendency for members of a population to work in unison, to get synchronized, is ubiquitous in nature, from flocks of birds to groups of neurons. It is a general property of our universe as much as gravity and electricity.

The US mathematician Norbert Wiener (the inventor of Cybernetics) believed that the activity of the brain is synchronized by a clock that is implemented by a group of neurons. Each neuron is a terrible clock, and is highly vulnerable. A group of neurons, though, each of them a bad clock, that influences each other, tends to converge towards a synchronized state that constitutes a much more reliable clock. Wiener proved mathematically that a population of interacting neurons (that adjust their frequency based on what their neighbors are doing) yields over time a population of synchronized neurons. He speculated that "frequency pulling" might be a universal method of self-organization for any complex system.

The US mathematician Charles Peskin ("Mathematical Aspects of Heart Physiology", 1975) reached a similar conclusion when studying the "clock" of the heart: its rhythm is set not by a specific cell but by the collective behavior of a group of cells. In both cases the advantage of having a population (rather than just one member) provide the clock is that the process becomes fault tolerant: if one fails, the whole corrects it; if one dies, the others are enough for the process to continue.

The US biologists David Welsh and Steve Reppert ("Individual Neurons Dissociated From Rat Suprachiasmatic Nucleus Express Independently Phased Circadian Firing Rhythms", 1995) discovered that the suprachiasmatic nucleus of the mammalian hypothalamus contains a circadian clock. This "clock" is actually made of a large population of independent, single-cell oscillators that collectively provide an accurate time keeping. This "master" clock is probably responsible for all the rhythms in the body of a mammal.

Populations of synchronized oscillators seem to be pervasive in nature. The Indian physicist Satyendra Bose discovered in 1924 that at low temperatures all bosons behave like one. Such a Bose-Einstein condensate was first achieved in a gas in 1995. This means that all their quantum waves are synchronized (or, better, "phase coherent").

The British physicist Brian Josephson ("Coupled Superconductors", 1963) discovered an odd quantum phenomenon called "Johnson junction", which is a consequence of quantum tunneling effect: an electric current arises between two weakly coupled superconductors (whose waves overlap slightly but don't interfere with each other too much) that are separated by a very thin non-conducting barrier. Richard Feynman realized that this was a special case of a universal phenomenon: the Josephson effect will occur for any pair of weakly coupled phase-coherent systems. It was later realized that the equations for the electrical oscillations in a Josephson junction are identical to the equations of the motion of a pendulum.

The US physicist Arthur Winfree studied the nonlinear equations of a population of coupled oscillators and verified that, if each oscillator can influence the others, the population as a whole has a tendency to get synchronized. The population does not need any leader in order to achieve this: it's the interaction among the various oscillators of the population that creates the order. Winfree realized that synchronization of a population of oscillators is a phenomenon similar to the phase transition of a substance (for example, water), in which all the molecules of that substance have to "cooperate" in order for the substance to change state (for example, to turn into ice). Organizing a population of oscillators is similar to organizing a population of molecules, except that the former gets organized in time and the latter gets organized in space.

The Japanese physicist Yoshiki Kuramoto proved that a system of such equations always has one obvious solution (the state of incoherence, in which the population is completely disorganized, a state which can actually be implemented in a very large number of ways) and sometimes also has another solution, one of complete synchronization. The latter occurs only when the initial states are not too chaotic. There is a threshold value for their chaos below which that solution of complete synchronization exists and above which it does not exist. The initial states have to be at least partially synchronized. Again, this principle evokes the threshold above or below which a phase transition occurs in a substance.

The US mathematician Steven Strogatz then proved that the state of incoherence is a state of unstable equilibrium, which means that it will sooner or later collapse into one of the other possible states, the states that represent synchronous behavior. In other words, Strogatz proved that synchronicity (and therefore self-organization) "will" emerge at some point in any system that exhibits partial synchronicity above the threshold.

Strogatz also figured out that, under certain circumstances, a population of coupled Josephson junctions will behave like Kuramoto's biological oscillators: the junctions will suddenly synchronize. Any system made of many independent oscillators that are weakly coupled, and that are coupled with the same intensity to all the others, will exhibit spontaneous synchrony.

The US physicist John Hopfield ("Neurons, Dynamics and Computation", 1994), the man who had rescued neural networks from oblivion, made the connection with self-organized systems: a system of synchronized oscillator is a self-organized system of the kind studied by Kauffman and Bak. Self-organization can be achieved in time or in space. Some interacting molecules, cells and atoms achieve self-organization in space through spontaneous reorganization, whereas some coupled oscillators achieve self-organization in time through spontaneous synchrony.

Synchronized oscillators exhibit mathematical properties that might explain natural phenomena. For example, chemical reactions can oscillate spontaneously. During such oscillations there is a point (the "phase singularity") that does not oscillate like the rest, in which the cycle amplitude collapses down to zero and its phase cannot be determined anymore. It turns out that this phase singularity generates a spiral wave that cannot be destroyed for as long as the phase singularity exists. The spiral is extremely resilient, almost invulnerable. These spiral waves emerge in chemical, biological and physical systems under the right conditions.

Phase transitions occur everywhere in nature. It is not only water that turns to ice or to steam. There are mental phenomena that exhibit the behavior of phase transitions. Since the time of the Swiss psychologist Jean Piaget, we have been studying how children go through different "stages" of understanding. Each stage emerges suddenly after a period of years during which no training and education would make it emerge. Stages still occur in the daily lives of adults. Sometimes it takes us time to understand a concept, especially a scientific one. We read the same sentences again, apparently incapable of grasping the meaning. Then suddenly we "get it". We understand what those sentences meant. From that point onwards, we will always understand it and it will be much easier to understand similar concepts. Something has happened in our brain that suddenly made it easier to do something that one second earlier was difficult. The brain has undergone a phase transition. And the functioning of the brain depends on synchronized oscillations of neurons.

### *The Emergence of a Science of Emergence*

Theories such as Prigogine's non-equilibrium Thermodynamics, Haken's Synergetics, Von Bertalanffy's General Systems Theory and Kauffman's complex adaptive systems all point to the same scenario: the origin of life from inorganic matter is due to nonlinear processes of self-organization. The same processes account for "emergent" phenomena at different levels in the organization of the universe, and, in particular, for cognition. Cognition appears to be a general property of systems, not an exclusive one of the human mind.

***Further Reading***
Bak, Per: HOW NATURE WORKS (Copernicus, 1999)
Buchler, Justus: METAPHYSICS OF NATURAL COMPLEXES (Columbia University Press, 1966)
Bunge Mario: TREATISE ON BASIC PHILOSOPHY (Reidel, 1974-83)
Cohen, Jack & Steward Ian: THE COLLAPSE OF CHAOS (Viking, 1994)
Coveney, Peter: FRONTIERS OF COMPLEXITY (Fawcett, 1995)
Dalenoort G.J.: THE PARADIGM OF SELF-ORGANIZATION (Gordon & Breach, 1989)
Dalenoort G.J.: THE PARADIGM OF SELF-ORGANIZATION II (Gordon & Breach, 1994)
**Davies, Paul: GOD AND THE NEW PHYSICS (Penguin, 1982)**
Eigen, Manfred & Schuster Peter: THE HYPERCYCLE (Springer Verlag, 1979)
Foerster, Heinz von: OBSERVING SYSTEMS (Intersystems, 1981)
Forrest, Stephanie: EMERGENT COMPUTATION (MIT Press, 1991)
Fuller, Buckminster: SYNERGETICS (Macmillan, 1975)
Fuller, Buckminster: COSMOGRAPHY ( Macmillan, 1992)
**Gell-Mann, Murray: THE QUARK AND THE JAGUAR (W.H.Freeman, 1994)**
Gleick, James: CHAOS (Viking, 1987)
Goldberg, David: GENETIC ALGORITHMS (Addison Wesley, 1989)
Haken, Hermann: SYNERGETICS (Springer-Verlag, 1977)
Hayles, Katherine: CHAOS BOUND (Cornell Univ Press, 1990)
**Kauffman, Stuart: THE ORIGINS OF ORDER (Oxford University Press, 1993)**
Kauffman, Stuart: AT HOME IN THE UNIVERSE (Oxford Univ Press, 1995)
**Koestler, Arthur: THE GHOST IN THE MACHINE (Henry Regnery, 1967)**
Kuramoto, Yoshiki: CHEMICAL OSCILLATIONS, WAVES, AND TURBULENCE (Springer-Verlag, 1984)
**Langton, Christopher: ARTIFICIAL LIFE (Addison-Wesley, 1989)**
Laszlo, Ervin: INTRODUCTION TO SYSTEMS PHILOSOPHY (Gordon & Breach, 1972)
Lewin, Roger: COMPLEXITY (Macmillan, 1992)
Lyapounov, Aleksander: The General Problem of the Stability of Motion (1892)
Mandelbrot, Benoit: THE FRACTAL GEOMETRY OF NATURE (W.H.Freeman, 1982)

Nicolis, Gregoire & Prigogine Ilya: SELF-ORGANIZATION IN NON-EQUILIBRIUM SYSTEMS (Wiley, 1977)
**Nicolis, Gregoire & Prigogine Ilya: EXPLORING COMPLEXITY (W.H.Freeman, 1989)**
Nicolis, Gregoire: INTRODUCTION TO NONLINEAR SCIENCE (Cambridge University Press, 1995)
Pattee, Howard: HIERARCHY THEORY (Braziller, 1973)
Prigogine, Ilya: INTRODUCTION TO THERMODYNAMICS OF IRREVERSIBLE PROCESSES (Interscience Publishers, 1961)
Prigogine, Ilya: NON-EQUILIBRIUM STATISTICAL MECHANICS (Interscience Publishers, 1962)
**Prigogine, Ilya & Stengers Isabelle: ORDER OUT OF CHAOS (Bantham, 1984)**
Salthe, Stanley: EVOLVING HIERARCHICAL SYSTEMS (Columbia University Press, 1985)
Salthe, Stanley: DEVELOPMENT AND EVOLUTION (MIT Press, 1993)
Strogatz, Steven: SYNC (Hyperion, 2003)
Thom, Rene': MATHEMATICAL MODELS OF MORPHOGENESIS (Horwood, 1983)
Thom, Rene': STRUCTURAL STABILITY AND MORPHOGENESIS (Benjamin, 1972)
Toffoli, Tommaso & Margolus, Norman: CELLULAR AUTOMATA MACHINES (MIT Press, 1987)
Varela, Francisco: PRINCIPLES OF BIOLOGICAL AUTONOMY (North Holland, 1979)
Von Bertalanffy, Ludwig: GENERAL SYSTEMS THEORY (Braziller, 1968)
Weiss, Paul: THE SCIENCE OF LIFE (1973)
Waldrop, Mitchell: COMPLEXITY (Simon & Schuster, 1992)
Winfree, Arthur: THE GEOMETRY OF BIOLOGICAL TIME (Springer-Verlag, 1980)
Wiener, Norbert: NONLINEAR PROBLEMS IN RANDOM THEORY (MIT Press, 1958)
Zeeman, Erich-Christian: CATASTROPHE THEORY (Addison-Wesley, 1977)

# THE NEW PHYSICS: THE UBIQUITOUS ASYMMETRY

## Physics and the Mind

Today's science, in particular Relativity and Quantum theories, present us with a world that is hardly the one we intuitively know. The reason is simple: Quantum and Relativity theories deal with the problems of the immensely large or the immensely small. Our brains were built to solve a different class of problems. They were built to deal with midsize, colored objects that move slowly in a three-dimensional space over a span of less than a century.

The vast majority of the theories of mind still assume that the world is a Newtonian world of objects, of continuous time, of absolute reality and of force-mediated causality. What that means is very simple: most theories of mind are based on a Physics that has been proven wrong. Newton's Physics does work in many cases, but today we know that it does not work in some other cases. We don't know whether mind belongs to the set of cases for which Newton's Physics is a valid approximation of reality, or whether mind belongs to the set of cases for which Newton's Physics yields wrong predictions. Any theory of mind that is based on Newton's Physics is a gamble.

For example, psychologists often like to separate the senses and feelings based on the intuitive fact that the senses gives us a photograph of the world, whereas pleasure and pain are the outcome of an interpretation of the world. When I see an object, I am transferring a piece of reality as it is inside my mind. When I feel pleasure, I am interpreting something and generating a feeling. This separation makes sense only if one assumes that objects do exist. Unfortunately, modern Physics has changed our perception of reality. What exists is a chaos of elementary particles, that our eyes "interpret" as an object. A chair is no more real than a feeling of pain. They are both created by my mind. Actually, what exists is truly waves of probabilities, that somehow our brain reduces to objects.

Modern Physics is not necessarily right (although Newton is necessarily wrong on several issues, otherwise Hiroshima would still be standing). But many theories of mind rely on a Physics that, de facto, is either Newton's or is a Physics that has not been invented yet.

## The Classical World: Utopia

Since we started with the assumption that our Physics is inadequate to explain at least one natural phenomenon, consciousness, and therefore cannot be "right" (or, at least, complete), it is worth taking a quick look at what Physics has to say about the universe that our consciousness inhabits.

Our view of the world we live in has undergone a dramatic change over the course of this century. Quantum Theory and Relativity Theory have changed the very essence of Physics, painting in front of us a completely different picture of how things happen and why they happen.

Let's first recapitulate the key concepts of classical Physics. Galileo laid them down in the 16th century. First of all, a body in free motion does not need any force to continue moving. Second, if a force is applied, then what will change is the acceleration, not the velocity (velocity will change as a consequence of acceleration changing). Third, all bodies fall with the same acceleration. A century later, Newton expressed these findings in the elegant form of differential calculus and immersed them in the elegant setting of Euclid's geometry. Three fundamental laws explain all of nature (at least, all that was known of nature at the time). The first one states that the acceleration of a body due to a force is inversely proportional to the body's "inertial" mass. The second one states that the gravitational attraction that a body is subject to is proportional to its "gravitational" mass. The third one indirectly states the conservation of energy: to every action there is always an equal reaction.

They are mostly rehashing of Galileo's ideas, but they state the exact mathematical relationships and assign numerical values to constants. They lent themselves to formal calculations because they were based on calculus and on geometry, both formal systems that allowed for exact deduction. By applying Newton's laws, one can derive the dynamic equation that mathematically describes the motion of a system: given the position and velocity at one time, the equations can determine the position and velocity at any later time. Newton's world was a deterministic machine, whose state at any time was a direct consequence of its state at a previous time. Two conservation laws were particularly effective in constraining the motion of systems: the conservation of momentum (momentum being velocity times mass) and the conservation of energy. No physical event can alter the overall value of, say, the energy: energy can change form, but ultimately it will always be there in the same amount.

In 1833 the Irish mathematician William Hamilton, building on the 1788 work of the Italian mathematician Luigi Lagrange (the trajectory of an object can be derived by finding the path which minimizes the "action", such action being basically the difference between the kinetic energy and the potential energy), realized something that Newton had only implied: that velocity, as well as position, determines the state of a system. He also realized that the key quantity is the overall energy of the system. By combining these intuitions, Hamilton redefined Newton's dynamic equation with two equations that derived from just one quantity (the Hamiltonian function, a measure of the total energy of the system), that replaced acceleration (a second-order derivative) with the first-order derivative of velocity, and that were symmetrical (once velocity was replaced by momentum). The bottom line was that position and velocity played the same role and therefore the state of the system could be viewed as described by six coordinates, the three coordinates of position plus the three coordinates of momentum. At every point in time one could compute the set of six coordinates and the sequence of such sets would be the history of the system in the world. One could then visualize the evolution of the system in a six-dimensional space, the "phase" space.

In the Nineteenth century two phenomena posed increasing problems for the Newtonian picture: gases and electromagnetism. Gases had been studied as collections of particles, but, a gas being made of many minuscule particles in very fast motion and in continuous interaction, this model soon revealed to be a gross approximation. The classical approach was quickly abandoned in favor of a stochastic approach, whereby what matters is the average behavior of a particle and all quantities that matter (from temperature to heat) are statistical quantities.

In the meantime, growing evidence was accumulating that electric bodies radiated invisible waves of energy through space, thereby creating electromagnetic fields that could interact with each other, and that light itself was but a particular case of an electromagnetic field. In the 1860s the British physicist James Maxwell expressed the properties of electromagnetic fields in a set of equations. These equations resemble the Hamiltonian equations in that they deal with first-order derivatives of the electric and magnetic intensities. Given the distribution of electric and magnetic charges at a time, Maxwell's equation can determine the distribution at any later time. The difference is that electric and magnetic intensities refer to waves, whereas position and momentum refer to particles. The number of coordinates needed to determine a wave is infinite, not six... As a by-product of his equations, Maxwell also discovered that light is an electromagnetic wave. Basically, Michael Faraday had shown that changes in magnetic fields produce electric fields; Maxwell realized that the opposite is also true: changes in electric fields cause magnetic fields. His equations describe this endless dance between electric and magnetic fields.

Philosophically speaking, these studies constituted a paradigm shift: reality consists not of discrete objects located in space but rather of a "field" distributed in space.

By then, it was already clear that Science was faced with a dilemma, one which was bound to become the theme of the rest of the century: there are electromagnetic forces that hold together particles in objects and there are gravitational forces that hold together objects in the universe, and these two forces are both inverse square forces (the intensity of the force is inversely proportional to the square of the distance), but the two quantities they act upon (electric charge and mass) behave in a completely different way, thereby leading to two completely different descriptions of the universe.

The inverse square law indirectly also implied that the "vacuum" has a role: the attraction decreases exponentially with distance; hence there is a relationship between a force and space. These two forces act at a distance, but somehow their "strength" depends on the distance, as if the "nothing" in between two bodies contributed to the measured strength at each side.

Newton's laws of motion apply to inertial frames and those laws are the same for all inertial frames. However, Maxwell noticed that an electric phenomenon in one inertial frame was a magnetic phenomenon in another. An electric phenomenon in movement gives rise to a magnetic phenomenon, and viceversa, but the "movement" depends on who is observing: if you observe the electric phenomenon while it is moving in front of you, you perceive a magnetic phenomenon; if you "ride" on the electrical phenomenon (which is therefore at rest from your viewpoint), you perceive instead only an electrical phenomenon. Maxwell's equations showed that the electromagnetic phenomenon "oscillates" like a wave and all such waves travel at 300 thousand kms/hour (in the vacuum). Light is one particular electromagnetic wave, hence that speed is now known as "the speed of light". What we perceive as different colors of light correspond to different frequencies, not velocities. And some frequencies of electromagnetic waves we can't perceive at all. The interaction between distant bodies does not happen instantaneously as Newton thought but is mediated by a "field".

Another catch hidden in all of these equations was that the beautiful and imposing architecture of Physics could not distinguish the past from the future, something that is obvious to all of us. All of Physics' equations were symmetrical in time. There is nothing in Newton's laws, in Hamilton's laws, in Maxwell's laws or even in Einstein's laws that can discriminate past from future. Physics was reversible in time, something that goes against our perception of the absolute and (alas) irrevocable flow of time.

### *The Removal of Consciousness*

In the process, something else had also happened, something of momentous importance, even though its consequences would not be appreciated for a few centuries. Galileo had fostered the mathematical study of nature: science is about creating a mathematical model of a natural phenomenon. René Descartes had introduced the "experimental method": science has to be based on experiments and proofs. Descartes started out by defining the domain of science. He distinguished between matter and mind, and decided that science had to occupy itself with matter. Therefore the schism was born that would influence the development of human knowledge for the next three centuries: science is the study of nature, and our consciousness does not belong to nature. Newton built on Galileo's foundations. Physics, in other words, had been forced to renounce consciousness and developed a sophisticated system of knowledge construction and verification that did not care about, and therefore did not apply to, consciousness. Scientists spoke of "nature" as if it included only inanimate, unconscious objects. No wonder that they ended up building a science that explains all known inanimate, unconscious phenomena, but not consciousness.

The Austrian physicist Erwin Schroedinger, one of the founders of Quantum Physics, identified two fundamental tenets of classical science: object and subject can be separated (i.e., the subject can look at the object as if it were a completely disconnected entity); and the subject is capable of knowing the object (i.e., the subject can look at the object in a way that creates a connection, one leading to knowledge). There is a subtle inconsistency in these two tenets: one denies any connection between subject and object, while the other one states an obvious connection between them.

### *Entropy: The Curse of Irreversibility*

The single biggest change in scientific thinking may have nothing to do with Relativity and Quantum theories: it may well be the discovery that some processes are not symmetric in time. Before the discovery of the second law of Thermodynamics, all laws were symmetric in time, and change could always be bi-directional. Any formula had an equal sign that meant one can switch the two sides at will. We could always replay the history of the universe backwards. Entropy changed all that.

Entropy was "discovered" around 1850 by the German physicist Rudolf Clausius in the process of revising the laws proposed by the French engineer Sadi Carnot, that would become the foundation of Thermodynamics. The first law of Thermodynamics is basically the law of conservation of energy: energy can never be created or destroyed, it can only be transformed. The second law (originally formulated by William Thompson "Kelvin" in 1852) states that any transformation has an energetic cost: this "cost" of transforming energy Clausius called "entropy" (which is numerically obtained by dividing heat by temperature). Natural processes generate entropy. Entropy explains why heat flows spontaneously from hot to cold bodies, but the opposite never occurs: "useful" energy can be lost in entropy, not viceversa. There can never be an isolated process that results in a transfer of energy from a cold body to a hotter body: it is just a feature of our universe. In a sense, entropy measures how useful energy is. Before entropy is created, all energy is useful (e.g., the explosion in the combustion engine). Afterwards, some energy has become largely useless (e.g. heat, noise, motion).

The first law talks about the quantity of energy, while the second law talks about the quality of such energy. Energy is always conserved, but something happens to it that causes it to "deteriorate". Entropy measures the amount of energy that has deteriorated (is not available anymore for further work).

Clausius summarized the situation like this: the energy of the universe is constant, the entropy of the universe is increasing.

In the 1870s, the German physicist Ludwig von Boltzmann deduced entropy from the motion of gas particles, i.e. from dynamic laws that are reversible in nature. Indirectly, Boltzmann proved that entropy (and therefore irreversibility) is an illusion, that matter at the microscopic level is fundamentally reversible. Convinced that bodies are made of a large number of elementary particles, Boltzmann used statistics and probability theory to summarize their behavior, since it would be impossible to describe each particle's motion and their innumerable interactions. He noticed that many different configurations (microstates) of those particles lead to the same external appearance (macrostate) of the system as a whole. For example, at the game of poker the macrostate of a "four of a kind" can be due to any hand that contains all four cards of one rank and any other card. Boltzmann realized that the molecules of an isolated gas tend to scatter until they are as randomly distributed as they can be, and that's the macrostate of thermodynamic equilibrium. This randomness turns out to correspond to Clausius' entropy, and this randomness can be expressed mathematically as the number of microstates that can create a macrostate. For example, a straight flush is the least random kind of poker hand because it can be generated by the smallest set of card distribution (it takes five sequential cards in the same suit). Boltzmann, living in the age of the slide rule, took the logarithm of this number because in this way the randomness of a combined macrostate is simply the sum of the randomness of the individual macrostates.

Boltzmann ended up with a statistical definition of entropy to characterize the fact that many different microscopic states of a system result in the same macroscopic state: the entropy of a macrostate is the logarithm of the number of microstates that can implement that macrostate. Intuitively, the law of entropy originates from a statistical trend: a system tends to evolve towards the macrostates with high entropy, i.e. macrostates that correspond to large numbers of microstates; basically from rare configurations towards more likely configurations. Entropy is a probability function, and the second law of Thermodynamics simply states that any system tends to drift towards the most probable state, which happens to be a state of higher entropy.

Boltzmann's implicit assumption was that every microstate is equally probable. The other implicit assumption of the second law is that the universe started in a state of low entropy. That creates the fundamental asymmetry that we recognize as the arrow of time: entropy tends to increase because it is a lot easier to increase than decrease, and that is because the beginning of the story was at low entropy. For example, we assume a low-entropy past when we trust our memories of it: our memories (including photographs, videos, diaries) could have been created in a myriad of ways, but the low-entropy explanation is that they reflect what really happened.

Boltzmann's definition emphasized that entropy turns out to be also a measure of "disorder" in a system: an ordered system has fewer microstates corresponding to a given macrostate. Entropy can also be interpreted as a measure of randomness, i.e. closed systems tend to drift from order to randomness.

Recast in terms of order and chaos/randomness, Thermodynamics was now a powerful intellectual tool that could be applied to other domains beyond heat engines.

The second law of Thermodynamics is an inequality: it states that entropy can never decrease. Indirectly, this law states that transformation processes cannot be run backward, cannot be "undone". Young people can age, but old people cannot rejuvenate. Buildings do not improve over the years: they decay. Scrambled eggs cannot be unscrambled and dissolved sugar cubes cannot be recomposed. The universe must evolve in the direction of higher and higher entropy. Some things are irreversible.

Newton's equations are symmetric in time, which means that theoretically the same process can run backwards. It is the second law of Thermodynamics which makes it impossible to go back in time, that introduces an "arrow" of time.

The universe as a whole is proceeding towards its unavoidable fate: the "heat death", i.e. the state of maximum entropy, in which no heat flow is possible, which means that temperature is constant everywhere, which means that there is no energy available to produce more heat, which means that all energy in the universe is in the form of heat. (An escape from the heat death would be possible if the energy in the universe were infinite).

Scientists were (and still are) puzzled by the fact that irreversibility (the law of entropy) had been deduced from reversibility (basically, Newton's laws). Mechanical phenomena tend to be reversible in time, whereas thermodynamic phenomena tend to be irreversible in time. Since a thermodynamic phenomenon is made of many mechanical phenomena, the paradox is how can an irreversible process arise from many reversible processes? It is weird that irreversibility should arise from the behavior of molecules which, if taken individually, obey physical laws that are reversible. We can keep track of the motion of each single particle in a gas, and then undo it. But we cannot undo the macroscopic consequence of the motion of thousands of such particles in a gas.

If one filmed the behavior of each particle of a gas as the gas moves from non-equilibrium to equilibrium, and then played the film backwards, the film would be perfectly consistent with the laws of Mechanics. In practice, though, systems never spontaneously move from equilibrium to non-equilibrium: the film is perfectly feasible, but in practice it is never made.

The only reason one could find was probabilistic, not mechanical: the probability of low-entropy macrostates is smaller, by definition, than the probability of high-entropy macrostates, so the universe tends to proceed towards higher entropy. Thus the second law seems to express the tendency of systems to transition from less probable states (states that can be realized by few microstates) to more probable states (states that can be realized by many microstates). Basically, there are more ways to be disorderly than to be orderly.

And one can rephrase the same idea in terms of equilibrium: since equilibrium states are states that correspond to the maximum number of microstates, it is unlikely that a system moves to a state of non-equilibrium, and likely that it moves to a state of equilibrium.

## *Recurrence*

The trick is that Boltzmann assumed that a gas (a discrete set of interacting molecules) can be considered as a continuum of points and, on top of that, that the particles can be considered independent of each other: if these arbitrary assumptions are dropped, no rigorous proof for the irreversibility of natural processes exists.

The French mathematician Henri Poincaré ("Sur le problème des trois corps et les équations de la dynamique", 1890), for example, proved just about the opposite: that every closed system must eventually revert in time to its initial state (the "recurrence theorem"). Thus everything that can happen "will" happen, and will happen infinite times. Poincaré proved eternal recurrence where Thermodynamics had just proved eternal doom. The German mathematician Ernst Zermelo immediately ("On a Theorem of Dynamics and the Mechanical Theory of Heat", 1896) noticed that this would violate the law of entropy, as the return to a previous state would imply that entropy at some point must decrease in order to return to its original value.

Boltzmann could find only one rational reply: that there might be universes in which entropy decreases to compensate for universes like ours in which entropy can never decrease.

It took the Belgian (but Russian-born) physicist and Nobel-prize winner Ilya Prigogine, in the 1970s, to provide a more credible explanation for the origin of irreversibility. He observed some inherent time asymmetry in chaotic processes at the microscopic level, which would cause entropy at the macroscopic level. He reached the intriguing conclusion that irreversibility originates from randomness which is inherent in nature.

Boltzmann's reformulation of the second law was probabilistic: it explained the entropy of the system as a property about a population of particles, not just one particle. The second law does not claim that every single particle is subject to it, but that closed systems (made of many particles) are subject to it. An individual particle may well be violating the second law for a few microseconds, but the millions of particles that make up a system will obey it (just like one person might win at the roulette once, but that episode does not change the statistical law that people lose money at the roulette). In 2002 Australian researchers, in fact, showed that microscopic systems may spontaneously become more orderly for short periods of time.

### *Disorder, Ignorance*

Entropy is a measure of disorder, and information is found in disorder (the more the microstates the more the information, ergo the more the disorder the more the information), so ultimately entropy is also a measure of information.

Later, several scientists interpreted entropy as a measure of ignorance about the microscopic state of a system, for example as a measure of the amount of information needed to specify it. Murray Gell-Mann summarized these arguments when he gave his explanation for the drift of the universe towards disorder. The reason that nature prefers disorder over order is that there are many more states of disorder than of order, therefore it is more probable that the system ends up in a state of disorder. In other words, the probability of disorder is much higher than the probability of spontaneous order, and that's why disorder happens more often than order.

Equilibrium states are also states of minimum information (a few parameters are enough to identify the state, e.g. one temperature value for the whole gas at a uniform temperature). Information is negative entropy and this equivalence would play a key role in applying entropy beyond Physics.

### *An Accelerated World*

Science has been long obsessed with acceleration. Galileo and Newton went down in history for managing to express that simple concept of acceleration. After them Physics assumed that an object is defined by its position, its velocity (i.e., the rate at which its position changes) and its acceleration (i.e., the rate at which its velocity changes). The question is: why stop there? Why don't we need the "rate an object changes its acceleration" and so forth? Position is a space coordinate. Velocity is the first derivative with respect to time of a space coordinate. Acceleration is the second derivative with respect to time of a space coordinate. Why do we only need two orders of derivative to identify an object, and not three or four or twenty-one?

Because the main force we have to deal with is gravity, and it only causes acceleration. We don't know any force that causes a change in acceleration, therefore we are not interested in higher orders of derivatives.

To be precise, forces are defined as things that cause acceleration, and only acceleration (as in Newton's famous equation "F=ma"). We don't even have a word for things that would cause a third derivative with respect to time of a space coordinate.

As a matter of fact, Newton explained acceleration by introducing gravity. In a sense Newton found more than just a law of Physics, he explained a millenary obsession: the reason mankind had been so interested in acceleration is that there is a force called gravity that drives the whole world. If gravity did not exist, we would probably never have bothered to study acceleration. Car manufacturers would just tell customers how long it takes for their car to reach such and such a speed. Acceleration would not even have a name.

### *Relativity: The Primacy of Light*

The Special Theory of Relativity was born ("On the Electrodynamics of Moving Bodies", 1905) out of Albert Einstein's belief that the laws of nature must be uniform, whether they describe the motion of bodies or the motion of electrons. Therefore, Newton's equations for the dynamics of bodies and Maxwell's equations for the dynamics of electromagnetic waves had to be unified in one set of equations. In addition, they must be the same in all frames of reference that are "inertial", i.e. whose relative speed is constant. Galileo had shown this to be true for Newton's mechanics, and Einstein wanted it to be true for Maxwell's electromagnetism as well. In order to do that, one must modify Newton's equations, as the Dutch physicist Hendrik Lorentz had already pointed out in 1892. The implications of this unification are momentous.

Relativity conceives all motion as "relative" to something. Newton's absolute motion, as the Moravian physicist Ernst Mach had pointed out over and over, is an oxymoron. Motion is always measured relative to something. Best case, one can single out a privileged frame of reference by using the stars as a meta-frame of reference. But even this privileged frame of reference (the "inertial" one) is still measured relative to something, i.e. to the stars. There is no frame of reference that is at rest, there is no "absolute" frame of reference. While this is what gave Relativity its name, much more "relativity" was hidden in the theory.

In Relativity, space and time are simply different dimensions of the same space-time continuum, as shown by the Russian mathematician Hermann Minkowski ("The Basic Equations for Electromagnetic Processes in Moving Bodies", 1908). Einstein showed that the length of an object and the duration of an event are relative to the observer. This is equivalent to calculating a trajectory in a four-dimensional spacetime that is absolute. The spacetime is the same for all reference frames and what changes is the component of time and space that is visible from your perspective. One person's time is another person's mixture of time and space.

All quantities are redefined in space-time and must have four dimensions. For example, energy is no longer a simple (monodimensional) value, and momentum is no longer a three-dimensional quantity: energy and momentum are one space-time quantity which has four dimensions. Which part of this quantity is energy and which part is momentum depends on the observer: different observers see different things depending on their state of motion, because, based on their state of motion, a four-dimensional quantity gets divided in different ways into an energy component and a momentum component. All quantities are decomposed into a time component and a space component, but how that occurs depends on the observer's state of motion.

This phenomenon is similar to looking at a building from one perspective or another: what we perceive as depth, width or height, depends on where we are looking from. An observer situated somewhere else will have a different perspective and will measure different depth, width and height. The same idea holds in space-time, except that now time is also one of the quantities that changes with "perspective" and the motion of the observer (rather than her position) determines what the "perspective" is. This accounts for bizarre distortions of space and time: as speed increases, lengths contract and time slows down (the first to propose that lengths must contract was, in 1889, the Irish physicist George Fitzgerald, but he was thinking of a physical contraction of the object, and Lorentz endorsed it because it gave Maxwell's equations a particularly elegant form, whether the observer was at rest or in motion). This phenomenon is negligible at slow speeds, but becomes very visible at speeds close to the speed of light.

An observer who travels away from a clock-tower at the speed of light, would always observe the same time, as if the clock's hands never moved and time was still. If the observer traveled at a speed slightly less than the speed of light, the observer would see the hands of the clock moving very slowly over the years as the light would take a long time to travel that distance. On the other hand an observer who travels very slowly away from the same clock-tower (all of us on human-made vehicles), would observe the clock's hands moving. Therefore time depends on the speed of the observer relative to the clock (or viceversa). A moment of time is slower at higher speed. Time intervals are dilated by higher speeds.

A further implication is that "now" becomes a meaningless concept: one observer's "now" is not another observer's "now". Two events may be simultaneous for one observer, while they may occur at different times for another observer: again, their perspective in space-time determines what they see. The traditional law of causality is an illusion. Two events that follow each other from an observer's point of view may be simultaneous from the point of view of another observer who is moving at a different speed. The present is a concept that depends on the observer. Each observer has a different set of contemporary events that constitute its present.

Even the very concept of the flow of time is questionable. There appears to be a fixed space-time, and the past determines the future. Actually, there seems to be no difference between past and future: again, it is just a matter of perspective.

Time and space complement each other: as one dilates, the other contracts. The traditional law of causality had ceased to exist, but a new sort of causality was introduced because any warping of space corresponded to a warping of time.

The speed of light is the same in every frame of reference. What changes in each frame of reference is the very notion of distance and duration. That's why the speed of light remains the same regardless of what the frame of reference is doing: what it is doing alters its space and time in such a way that the speed of light remains the same for everybody.

Mass and energy are not exempted from "relativity". The mass and the energy of an object increase as the object speeds up. This principle violates the traditional principle of conservation, which held that nothing can be destroyed or created, but Einstein proved that mass and energy can transform into each other according to his famous formula $E=mc^2$: a particle at rest has an energy equal to its mass times the speed of light squared. (Note that the equation does not apply to things like photons that are not quite "objects" and that are never "at rest"). A very tiny piece of matter can release huge amounts of energy. Scientists were already familiar with a phenomenon in which mass seemed to disappear and correspondingly energy seemed to appear: radioactivity, discovered in 1896. But Einstein's conclusion that all matter is energy was far more reaching.

Light has a privileged status in Relativity Theory. The reason is that the speed of light is always the same, no matter what. If one runs at the speed of a train, one sees the train as standing still. On the contrary, if one could run at the speed of light, one would still see light moving at the speed of light. Most of Relativity's bizarre properties are actually consequences of this postulate. Einstein had to adopt the Lorentz transformations of coordinates, which leave the speed of light constant in all frames of reference, regardless of the speed it is moving at, but, in order to achieve this result, one must postulate that moving bodies contract and moving clocks slow down by an amount that depends on their speed.

If all this sounds unrealistic, remember that according to traditional Physics the bomb dropped on Hiroshima should have simply bounced, whereas according to Einstein's Relativity it had to explode and generate a lot of energy. That bomb remains the most remarkable proof of Einstein's Relativity.

Note that most of the equations in Einstein's Relativity had already been derived by others (Lorentz, Fitzgerald, Poincare) but they were merely attempts at explaining experimental data. Einstein introduced the principle of relativity (that the laws of physics must be the same in all inertial systems), and those equations became natural consequences. De facto, Einstein made a metaphysical choice: he decided that space and time are not absolute. Once the metaphysics changed, the oddities of the experimental data went away, or, better, became the natural consequences of a bigger oddity.

## *Life On A World Line*

The speed of light is finite and one of Relativity's fundamental principles is that nothing can travel faster than light. As a consequence, an object located in a specific point at a specific time will never be able to reach space-time areas of the universe that would require traveling faster than light.

The "light cone" of a space-time point is the set of all points that can be reached by all possible light rays passing through that point. Because the speed of light is finite, that four-dimensional region has the shape of a cone (if the axis for time is perpendicular to the axes for the three spatial dimensions). The light cone represents the potential future of the point: these are all the points that can be reached in the future traveling at the speed of light or slower. By projecting the cone backwards, one gets the light cone for the past. The actual past of the point is contained in the past light cone and the actual future of the point is contained in the future light cone. What is outside the two cones is unreachable to that point. And, viceversa, no event located outside the light cone can influence the future of that point. The "event horizon" of an observer is a space-time surface that divides space-time into regions which can communicate with the observer and regions which cannot.

The "world line" is the spatiotemporal path that an object is actually traveling through space-time. That line is always contained inside the light cone.

Besides the traditional quantity of time, Relativity Theory introduces another type of time. "Proper" time is the space-time distance between two points on a world line, because that distance turns out to be the time experienced by an observer traveling along that world line.

Relativity erased the concept of an absolute Time, but in doing so it established an even stronger type of determinism. It feels like our lives are rigidly determined and our task in this universe is simply to cruise on our world line. There is no provision in Relativity for free will.

### *General Relativity: The Illusion of Gravity*
Newton explained how gravity works, but not what it is.

Einstein's Gerneral Relativity Theory ("The Foundation of the Generalised Theory of Relativity", 1916) is ultimately about the nature of gravitation, which is the force holding together the universe. Relativity explains gravitation in terms of curved space-time, i.e. in terms of geometry.

The fundamental principle of this theory (the "principle of equivalence") is actually quite simple: any referential frame in free fall is equivalent to an inertial reference frame. That is because if you are in a free fall, you cannot perceive your own weight, i.e. gravity. Gravity is canceled in a frame of reference that is free falling, just like the speed of an object is canceled in a frame of reference which is moving at the same speed.

Newton had introduced two kinds of mass: the gravitational mass and the inertial mass. The gravitational mass of a body determines the power of gravitational attraction. The inertial mass of a body determines how strong a force has to be to move that body, or, conversely, how fast the body will accelerate when subject to a force. In theory, the two masses could be completely different. In practice, they are the same. That's why a uniform acceleration and standing still in a gravitational field turn out to have the same effect. This means that the acceleration generated by a gravitational field does not depend on the mass, as Galileo famously proved dropping balls of different material from the Leaning Tower of Pisa. This is unique to gravitation: in the case of any other force, the mass of the body determines the acceleration of the body.

Since gravity is canceled when you are in free fall, the world around you will still behave the way it behaves in an inertial frame of reference. For example, if an object is falling while you are falling, the object will appear to be standing still.

If you can't see what is going on outside, and all you can measure is the 9.8 $m/sec^2$ acceleration of an object that you let fall, you can't decide whether you are standing still, and subject to Earth's gravity, or you are accelerating in empty space. All you observe is an acceleration of 9.8. If you are still, that's the acceleration you expect for any falling object. If you are in a rocket that is accelerating upwards at 9.8, that's the acceleration you expect for any object left behind. The effect is the same. Therefore, Einstein concluded, you can treat them as one: gravity and acceleration are equivalent.

This, however, is true only "locally". The Earth and Mars are both attracted to the Sun but the direction is usually different. Globally, the acceleration required to cancel gravity is different in different places. This means that each point in space must have its own "distortion" so that the principle of equivalence still holds. And therefore gravity must be a property of spacetime.

## *The Principle of Equivalence Revisited*

Einstein enunciated the principle of equivalence like this: no local experiment can detect the presence of a gravitational field. We cannot determine whether we are being accelerated in the absence of gravitational fields or we are at rest in the middle of a gravitational field. He knew that this statement is true only if one considers small regions of space and over brief intervals of time. Einstein concluded that gravitation is not really a force: it is a side effect of a different phenomenon.

Free fall (rather than stillness and rather than uniform speed) is the natural state of motion. After all, all objects fall the same way into a gravitational field. An object in free-fall towards a bigger body is actuality moving by inertia. The reason it looks like it is accelerating is that its time scale stretches at an accelerated rate as the object approaches the bigger body. In reality, the falling object is not accelerating at all. There is no force at work. It is just the time scale that changes. The change is in the geometry spacetime, not in the forces operating on the object.

The principle of equivalence basically states that there is no difference between an object in free-fall in a gravitational field and an object accelerating far away from any gravitational field.

This principle of equivalence is obviously false, if nothing else because of the tidal effects of a gravitational field that do not exist for an accelerating object far away from gravitational fields. But it also false simply because there is no such place without gravitational fields: it is just a matter of how small a field we are willing to measure. The key word is "local": no local experiment can detect the presence of a gravitational field; but the definition of "local" seems arbitrary, and, in general, nothing is truly "local". Einstein's principle is true only in an infinitely small region of spacetime. (In fact, it is debatable whether Einstein's very General Relativity complies with the principle of equivalence that inspired it).

However the outcome of the principle of equivalence was the equivalence between gravitation and acceleration.

When viewed from the broader canvas of spacetime, a change in velocity is but a change in orientation, and an acceleration is equivalent to a rotation in spacetime.

The principle of equivalence was just the thought experiment that helped realize how gravitation is not really a force like electromagnetism but has to do with spacetime itself.

## *Curving Spacetime*

Since gravitation is natural motion, Einstein's idea was to regard free falls as natural motions, i.e. as straight lines in spacetime. The only way to achieve this was to assume that the effect of a gravitational field is to produce a curvature of space-time: the straight line becomes a "geodesic", the shortest route between two points on a warped surface (if the surface is flat, then the geodesic is a straight line). Bodies not subject to forces other than a gravitational field move along geodesics of space-time.

The curvature of space-time is measured by a "curvature tensor" originally introduced in 1854 by the German mathematician Bernhardt Riemann. The Riemann geometry comprises the classical Euclidean geometry as a special case, but it is much more general.

Minkowski's four-dimensional spacetime is characterized by a "metrics". A metrics is a 4x4 matrix, each row and column representing one of the dimensions. The metrics for Newton's spacetime has zeros everywhere except in the diagonal of the matrix. The diagonal has values 1,1,1 and -1. This means that Pythagoras' theorem still works, and time is an added dimension. The zeros in the other positions of the matrix specify that the space is flat. When the ones and the zeros change, their values specify a curvature for spacetime. Euclidean geometry works only with the flat-space metrics. Riemann's geometry works with any combination of values, i.e. with any degree and type of curvature.

A specific consequence of Riemann's geometry is that "force" becomes an effect of the geometry of space. A "force" is simply the manifestation of a distortion in the geometry of space. Wherever there is a distortion, a moving object feels a "force" affecting its motion. Riemann's geometry is based on the notion of a "metric (or curvature) tensor", that expresses the curvature of space. On a two-dimensional surface each point is described by three numbers. In a four-dimensional world, it takes ten numbers at each point. This is the metric tensor. Euclid's geometry corresponds to one of the infinite possible metric tensors (the one that represents zero curvature).

Not only space and time are relative, but space-time is warped.

With his field equations, Einstein made the connection with the physical world: he related the curvature of space-time caused by an object to the energy and momentum of the object (precisely, the curvature tensor to the "energy-momentum tensor"). Einstein therefore introduced two innovative ideas: the first is that we should consider space and time together (three spatial dimensions and one time dimension), not as separate; the second is that what causes the warps in this space-time (i.e., what alters the metric from Euclid's geometry) is mass. A mass does not voluntarily cause gravitational effects: a mass first deforms space-time and that warping will affect the motion of other objects that will therefore be indirectly feeling the "gravitational force" of that mass.

The mass also has an effect on the "time" part of space-time: clocks in stronger gravitational fields (bigger warp) slow down compared with clocks in weaker gravitational fields (smaller warp).

Summarizing: the dynamics of matter is determined by the geometry of space-time, and that geometry is in turn determined by the distribution of matter. Space-time acts like an intermediary device that relays the existence of matter to other matter.

There is an analogy with Maxwell's theory of electromagnetism. If one thinks of the "metrics" as a metric field, then the metric field bends the trajectory of bodies that have mass-energy the same way that the electromagnetic field bends the trajectory of bodies that have electric charges.

Incidentally, this implies that mass-less things are also affected by gravitation. This includes light itself: a light beam is bent by a gravitational field. Light beams follow geodesics, which may be bent by a space-time warp.

### *Relativity and Common Sense*
Special Relativity asked the laws of nature be the same in all inertial frames; which implied that they had to be invariant with respect to the Lorentz transformations. As a consequence, Einstein had to accept that clocks slow down and bodies contract. With General Relativity he wanted laws of nature to be the same in all frames, inertial or not (his field equations basically removed the need for inertial frames). This implies that the laws of nature must be "covariant" (basically must have the same form) with respect to a generic transformation of coordinates. That turned out to imply a further erosion of the concept of Time: it turned out that clocks slow down just for being near a gravitational field.

Furthermore, one of the holy laws of Science, the conservation of energy, makes no sense in General Relativity: it is not clear how to measure "gravitational energy" (which in Newton's Physics was easily calculated) and it is not clear what "conserving" means in a spacetime in which time too is warped.

While apparent paradoxes (such as the twins paradox) have been widely publicized, Relativity Theory has been amazingly accurate in its predictions and so far no serious blow has been dealt to its foundations. While ordinary people may be reluctant to think of curved spaces and time dilatations, all these phenomena have been corroborated over and over by countless experiments.

### *Twins*
The paradox of the twins (devised by Einstein in person) is due to the fact that... everything is relative. If a twin leaves the Earth, travels to another planet with a speed close to the speed of light, and comes back to the Earth, this twin will be younger than the one that stayed on Earth. The reason is that clocks slow down as speed increases (time dilation).

However, according to Relativity, one can also run the experiment the other way around: from the point of view of the twin that departs the Earth, it is the Earth that travels away and then comes back. In this case, the twin who travels at high speed, and therefore who is younger, is the twin who stayed on the Earth. Thus the second twin is younger if measured from the first twin, but the first twin is younger if measured from the second twin: these measurements cannot both be true at the same time. Depending on which reference frame you use, you get two contradictory results.

Einstein solved the paradox by pointing out that the two situations are not symmetric. The twin who leaves the Earth has to apply an acceleration to get out of the Earth; then decelerate, turn and accelerate again to return to the Earth. All of this violates the principle of Relativity: the twin that departed the Earth has done something absolute.

Even if one assumes that the twin does not accelerate and decelerate, the fact remains that it changes direction. In a sense, there are three (not just two) inertial frames: the twin that stays on Earth, the twin that travels away from the Earth, and the twin that travels towards the Earth. Thus the elapsed time for the first twin is calculated by adding up two motions referred to the same frame (the Earth), whereas the elapsed time for the second twin must be calculated by adding up two motions referred to two different frames (the one moving away from the Earth and the one moving towards the Earth). Thus there is an absolute difference between the first twin measuring the second twin and the second twin measuring the first twin. The twin who becomes younger is the one leaving the Earth.

That said, it is important to remember that this "becoming younger" has nothing to do with bodily aging: it is only referred to time measured by clocks.

You can in fact dream up several "paradoxes" based on the same idea of going back and forth. Imagine, for example, that i cut a 1 cm circular hole from a sheet of paper. Now i move the sheet of paper far away, and move the circular piece high up in the air. Then i move the sheet of paper at very high speed towards the point where it will meet the circular piece that i am letting fall at a point in time such that it perfectly hits the hole. From the point of view of the circular piece, the sheet of paper is traveling at a very high speed, therefore it is shrinking, and, in particular, the hole in the middle is shrinking: therefore the circular piece will no longer be able to go through the hole. From the point of view of the sheet of paper, it is the circular piece that is traveling at very high speed, and thus shrinking: therefore the circular piece will easily pass through the hole. Imagine if instead of paper, you used spaceships: depending on which reference frame you use, the spaceships collide or they smoothly pass each other. This is not just a detail.

The solution of this second paradox is similar to the first one: we have done something at the very beginning, i.e. moving the sheet of paper far away. No matter how slowly we did that, we caused a change in its size relative to the circular hole (and viceversa). Thus, when we start moving the sheet of paper in the opposite direction, we cannot use its original size to compute the shrinking. When the sheet of paper and the circular piece meet, they are again the exact same size that they were at the beginning of the experiment.

### *Relativistic Cosmology*
Einstein's equations described more than just the interaction between two bodies (like Newton's gravitational equations did): they described the very story of the universe. One could play that film backwards or forward, and derive how the universe used to be or what it will be like.

Einstein briefly toyed with the idea of a "cosmological constant". He was not happy to discover that his equations predicted a universe in continuous turmoil (and most likely doomed to collapse under the effect of gravitation), so he introduced a constant in his equations to counterbalance gravitation (basically, a sort of "anti-gravity") and make the universe static. In 1922, however, Alexander Friedmann proved that his original equations simply described a universe that is not static and, in fact, expands uniformly. When in 1929 Edwin Hubble showed that the universe is expanding, Einstein realized that the turmoil was real and decided that there was no need for his cosmological constant.

Density of mass plays a crucial role in Einstein's equations: the denser the mass, the bigger the warp it causes to space-time, the stronger the gravitational effect felt by nearby matter. Thus collapsing stars are particularly relevant objects in Einstein's universe. In 1967, the first "pulsar" was observed: a rapidly-spinning collapsed star.

Shortly after Einstein published his gravitational field equation, in 1916 the German physicist Karl Schwarzschild found a solution that determines the gravitational field for any object, given its mass and its size. That solution goes to infinity for a specific ratio between mass and size: basically, if the object is dense enough (lots of mass in a tiny size), the gravitational attraction it generates is infinite. Nothing, not even light, can escape this object, which was therefore named "black hole" (by John Wheeler). And everything that gets near it is doomed to fall into it, and be trapped in it forever.

### *The General is not Special*
General Relativity actually violates some of the laws of Special Relativity, although it has a good excuse: spacetime is dynamic. The total energy of particles moving through a space that is changing is not conserved. If you still want to use the old-fashioned Newtonian concept of "the energy of the gravitational field", then energy is conserved, but in Einstein's picture the gravitational field is just geometry, not really energy.

The second major contradiction between Special and General Relativity is that in the latter objects can move apart at a speed greater than the speed of light. For example, if the universe is expanding, two galaxies can be observed to be moving apart faster than "c" by us. In general, gravity causes spacetime to warp which causes clocks to run slower, and therefore the speed of light at a certain location as measured by an observer from a different location could be different than "c". Einstein's Special Relativity constrains velocity at a maximum of "c" for systems that are inertial, but gravity causes (a spacetime warp that causes) acceleration, which means that the systems of reference are no longer inertial, which means that the speed of light observed in the other system can be faster or slower than "c". If we were freefalling into a black hole, we would measure the speed of light slower when looking towards the black hole and faster when looking away from the black hole.

### *Quantum Theory: The Wave*

Quantum Theory was the logical consequence of three discoveries. In 1900 the German physicist Max Planck solved the mystery of radiation emitted by heated objects (that Newton's physics failed to explain): he realized that atoms can emit energy only in discrete amounts. Nature seemed to forbid exchanges of energy in between those discrete values. According to Planck, therefore, the energy of light is proportional to the frequency: $E=h\nu$ (where "h" is Plack's constant).

The radioactive decay law, discovered in 1902 by Ernest Rutherford and Frederick Soddy, does not permit the determination of the time and trajectories of the particles emitted by a radioactive atomic nucleus. The nucleus seems to break up "spontaneously", at an unpredictable time. The only thing we can know is the nucleus' "half-life": how long it will take for half the nuclei in a sample to decay. For example, if the half life of a nucleus were 1,000 years, half the nuclei of that kind will decay within 1,000 years, and then half of the remaining in the next 1,000 years, and so forth.

In 1905 Einstein, to explain the photoelectric effect, argued that light must be physically made of packets ("photons") whose energy is proportional to the frequence.

In 1913 the Danish physicist Niels Bohr solved another mystery, the structure of the atom: electrons turn around the nucleus and are permitted to occupy only some orbits (or, better, the angular momentum of an electron occurs only in integer multiples of a constant, which happens to be proportional to Planck's constant). Again, Nature seemed to forbid existence in between orbits. The electron "jumps" from one orbit to another orbit without ever being in the space in between the two orbits (as if it stopped existing in the old orbit and was suddenly created again in the next orbit).

In 1925 George Uhlenbeck and Samuel Goudsmit discovered that each electron "spins" with an angular momentum of one half Planck constant.

(Particles actually don't spin, but interact as if they were spinning, hence the property that defines how they interact is called "spin" and particles are said to be "spinning"). The "spin" does not vary: the electron always rotates with the same "spin". It would turn out that every particle has its own spin, and the spin for any kind of particle is always the same.

The fundamental assumption of Quantum Theory is that any field of force manifests itself in the form of discrete particles (or "quanta"). Forces are manifestations of exchanges of discrete amounts of energy. For example, electromagnetic waves carry an energy which is an integer multiple of a fundamental constant, the "Planck constant".

A way to solve the apparent paradox of Bohr's electrons was discovered by the French physicist Louis de Broglie ("Waves and Quanta", 1923) after Einstein had made the same assumption regarding light: if an electron is viewed as a wave spreading over many orbits, the electron does not need to "jump" from one orbit to another. The electron "is" in all orbits at the same time, to some degree. De Broglie proved that the equation for a standing wave matched the behavior of the electron. Each particle is associated with a wave whose wavelength is inversely proportional the particle's momentum. That equation expressed a relationship between quantities of matter (e.g., speed, momentum, energy) and quantities of waves (e.g., wavelength and frequency). Thus he concluded that waves and particles are dual aspects of the same phenomena: every particle behaves like a wave. One can talk of energy and mass (quantities previously associated only to matter), or one can talk of frequency and wavelength (quantities previously associated only to waves). The two descriptions are equivalent, or, better, one complements the other. It didn't take long to observe "interference patterns" (typical of waves) among streams of electrons, and therefore confirm de Broglie's theory. Einstein's Relativity had shown that energy and matter were dual aspects of the same substance. De Broglie showed that particles and waves were dual aspects of the same phenomenon.

The character of this relationship was defined by Werner Heisenberg in Germany ("Quantum-Theoretical Re-interpretation of Kinematic and Mechanical Relations", 1925) and Erwin Schroedinger in Austria ("An Undulatory Theory of the Mechanics of Atoms and Molecules", 1926). Both devised equations that replaced the equations of Newton's physics, but both equations had unpleasant consequences: Heisenberg's equation (based on matrix algebra) implied that the result of a physical experiment depends on the order in which the calculations were performed, and Schroedinger's equation (based on wave mechanics) implied that each particle could only really be considered a wave. Schroedinger wanted to remove the discrete jumps (that were inherent in Heisenberg's formulation) and restore the continuum of classical Physics. His equation, after all, simply replaces Newton's (or, better, Hamilton's) equations and predicts the state of the system at a later time given the current state; except that his "system" is not a confined object but a wave. He thought of the wave as describing the location of the object (i.e., the object being spread out in space). However, experiments showed that the object (e.g., the electron) was a very confined object (just like in classical Physics) while Schroedinger's equation described it as a wave spread out in space. Max Born ("On the quantum mechanics of collisions", 1926) realized the implications of the wave-particle duality: the wave associated to a particle turns out to be a "wave of probabilities", that accounts for the alternative possibilities that open up for the future of a particle. In other words, the wave summarizes the possible values for the electron's attributes (e.g., position, energy, spin) and how those values may evolve over time (the square of the wave's amplitude represents the probability of finding a given value for an attribute when measuring that attribute). In particular, Schroedinger's wave is not a representation of where the object is spread out but of all the places where the object could possibly be, each to a certain degree of probability. This meant that the position of a particle cannot be known for sure: we can only guess it from a distribution of probability. We only know the probability of finding a particle in a given position.

The state of a particle is described by this "wave function" which summarizes (and superposes) all the alternatives and their probabilities. The wave function contains all the information there is about the particle (or, in general, about a system). It contains the answers to all the questions that can be asked about the particle.

The reason this is a "wave" of probabilities and not just a set of probabilities is that Schroedinger's equation that describes it is the equation of an electromagnetic wave.

Schroedinger's equation describes how this wave function evolves in time, and is therefore the quantum equivalent of Hamilton's equations. The Schroedinger equation fixes, deterministically, the temporal development of the state of the universe. But at every point in time the wave function describes a set of possibilities, not just one actuality. The particle's current state is actually to be thought of as a "superposition" of all those alternatives that are made possible by its wavelike behavior. A particle's current state is, therefore, a number of states: one can view the particle as being in all of those states at the same time. This is a direct consequence of a particle not being just a particle but being also a wave.

John von Neumann realized that, mathematically speaking, a classical system is represented in Newton's Physics by a point in a six-dimensional phase space (three coordinates for the position and three for the velocity), whereas quantum systems are represented by vectors in a vector space.

As Born put it, the motion of particles follows the law of probabilities, but the probability itself follows the law of causality.

In 1927 Bohr stated the ultimate paradox of the wave-particle duality: everything is both particle and wave, but one must choose whether to measure one or the other aspect of nature, and then stick to it. If you try to mix the two, you run into contradictions.

A particle is described by a function of probabilities. When it is observed by an instrument, the function "collapses" to one specific value. The transition from the world of the wave to the world of the particle is traumatic. The measurement that collapses the wave function creates an irreversible arrow of time. The fact that a measurement causes the collapse of the wave function (also called "state-vector reduction") is de facto a natural law that has to be added to the classical ones.

In Thermodynamics the microscopic laws of Physics were still Newtonian and therefore reversible. In Quantum Mechanics the microscopic laws of Physics are already irreversible, because nothing can undo the measurement: once you measure the position or the momentum of a particle, you have forever changed the state of the universe.

## *The Planck Constant*

Of course, one explanation begs another: introducing the Planck constant helps explain phenomena that Newton could not explain, but the mystery now is the Planck constant itself: what is it, what does it represent? Newton's physics (as well as Einstein's physics) assumed that the most fundamental units of the universe were the point and the instant. Quantum Theory introduces a fundamental unit that is bigger than a point and an instant, and seems to be as arbitrarily finite as Newton's points were infinitesimally small. Unlike Newton's points and instants, that have no size, the Planck constant has a size: a length, height and width of $10^{-33}$ centimeters and a time interval of $10^{-43}$ seconds.

Einstein had warped space and time, but Quantum Theory did worse: it turned them into grids. (One could even argue that the very notion of measuring a distance such as "$10^{-33}$ centimeters" depends on Newton's concept of space, and thus we don't quite know what we mean when we say that Planck's length is "$10^{-33}$ centimeters").

## *Enter Uncertainty*

In classical Physics, a quantity (such as the position or the mass) is both an attribute of the state of the system and an observable (a quantity that can be measured by an observer). Quantum Theory makes a sharp distinction between states and observables. If the system is in a given state, an observable can assume a range of values (so called "eigenvalues"), each one with a given probability. The evolution over time of a system can be viewed as due (according to Heisenberg) to time evolution of the observables or (according to Schroedinger) to time evolution of the states.

An observer can measure at the same time only observables that are compatible. If the observables are not compatible, they stand in a relation of mutual indeterminacy: the more accurate a measurement of the one, the less accurate the measurement of the other. Position and momentum are, for example, incompatible. This is a direct consequence of the wave-particle dualism: only one of the two natures is "visible" at each time. One can choose which one to observe (whether the particle, that has a position, or the wave, that has a momentum), but cannot observe both aspects at the same time.

Precisely, Heisenberg's famous "uncertainty principle" (or, better, "indeterminacy principle") states that there is a limit to the precision with which we can measure, at the same time, certain pairs of quantities, notably the momentum and the position of a particle. If one measures the momentum, then it cannot measure the position, and viceversa. Technically speaking: the product of uncertainties in position and in momentum must be greater than Planck's constant. This is actually a direct consequence of Einstein's equation that related the wavelength and the momentum (or the frequency and the energy) of a light wave: if coordinates (wavelength) and momentum are related, they are no longer independent quantities. Einstein never believed in this principle, but he was indirectly the one who discovered it.

A similar principle applies to other incompatible observables, for example between time and energy: one cannot measure energy precisely at a precise instant in time. Either the time or the amount of energy has to be imprecise. Hence, in theory, violations of energy conservation can occur... but we cannot observe them. The more the energy missing (unaccounted for), the faster it will be returned (the shorter the period of time before it is accounted for). If you try to measure energy at a precise time, then no information is known on how much energy is there.

Bohr summarized this oddity in his principle of complementarity ("The Quantum Postulate and the Recent Development of Atomic Theory", 1928): we are limited in our understanding of nature by pairs of inherently indeterminate quantities, a fact which is ultimately a consequence of the particle-wave duality. The more precise a member of the pair, the less precise its partner.

The wave function contains the answers to all the questions that can be asked about a system, but not all those questions can be asked simultaneously. If they are asked simultaneously, the replies will not be precise.

The degree of uncertainty is proportional to the Planck constant. This implies that there is a limit to how small a physical system can be, because, below a quantity proportional to the Planck constant and called "Planck length", the physical laws of Quantum Theory stop working altogether. The Planck scale ($10^{-33}$ cm, i.e. the shortest possible length, and

$10^{-43}$ sec, i.e. the time it takes for a light beam to cross the Planck length, i.e. the shortest possible time tick) is the scale at which space-time is no longer a continuum but becomes a grid of events separated by the Planck distance. What happens within a single cell of the grid, is beyond the comprehension of Physics. As the US physicist John Wheeler suggested in the 1950s, even the very notions of space and time stop making sense in this "quantum foam".

Note that Heisenberg does not forbid precise measurements of "compatible observables", for example of position, charge and spin. It only applies to "incompatible observables", which are couples: position/momentum, energy/time, electric field/magnetic field, angle/angular momentum, etc.

The uncertainty predicted by Quantum Theory (and verified by countless experiments in countless laboratories) has been sometimes interpreted as a consequence of the fact that, at the microscopic level, one cannot pretend that a measurement is "objective" at all: a measurement is an interaction between two systems, which, like all interactions, affects both. But that is not quite where Heisenberg's calculations came from. They originate, as everything else, from Planck's constant.

For the record, there had been other "principles of uncertainty" in Physics, and an important one in Mathematics, the one discovered by Joseph Fourier in the 19th century that a signal cannot be simultaneously localized both in time and in frequency: for example, there is a limit to the precision of the simultaneous measurement of the duration and frequency of a sound.

### *Zero-Point Energy*

As a consequence of quantum uncertainty, Planck and Heisenberg proved that at that scale, the vacuum of empty space is actually "full" of all sorts of subtle events. If you remove all particles, i.e. you have absolute certainty about the position of particles, then you know nothing about momentum and therefore about energy. There has to be a minimum amount of energy: the zero-point energy. Or, better: kinetic and potential energy cannot both be zero at the same time. If one is zero, the other one cannot be zero. The total is therefore always more than zero, even in a vacuum with no particles.

In 1930 Paul Dirac speculated that the all-pervading vacuum must actually be filled with particles that continuously appear and disappear in a random way. They live very short lives (limited by Heisenberg's principle), but the total effect of their brief existence is a fluctuation of energy in the vacuum. Thus the "vacuum" is not empty at all, and it actually generates some energy. In fact, one could view the vacuum as a warehouse where all possible particles have been stored, and occasionally they may become "real" (i.e., observable).

The way to create the vacuum is to lower the temperature to the absolute zero: this gets rid of any radiation. The energy produced by the vacuum is thus called the "zero-point energy". For example, helium near the absolute zero does not freeze, because the vacuum "warms it up" with its zero-point energy. The Dutch physicist Hendrik Casimir even showed how this all-pervading zero-point energy could be detected ("On The Attraction Between Two Perfectly Conducting Plates," 1948); and, in 1996, was finally detected what is now known as the "Casimir effect". The zero-point energy itself cannot be measured precisely, but a "change" in the zero-point energy can be measured precisely. The Casimir effect turns out to be inversely proportional to the fourth power of the distance.

Energy is everywhere, and, therefore, potentially, there is a lot of it. Alas, that energy cannot be used because "using" it would reduce the amount of the zero-point energy, which is, by definition, impossible.

This was the culmination of the eccentricities of Quantum Theory: that the vacuum was not empty.

Thus Quantum Theory predicts that the universe exists on a grid of spacetime values, and that there is something within the elements of this grid, something that does not quite belong to the universe (or, at least, does not belong to Quantum Physics) but has nonetheless an energy that can interact with the universe (be detected by people living on the grid of our universe).

### *Mass*

Galileo "discovered" inertia: bodies that are at rest tend to remain at rest, and bodies that are moving tend to continue moving at the same speed in the same direction, unless a force is applied. Newton turned Galileo's inertia into a quantitative property of matter: mass. Newton showed that mass was the object of forces, and the effect of forces on mass was to accelerate it. Forces and accelerations were visible entities. Mass was an invisible property of matter.

Newton's mass was three things in one: it was resistance to acceleration, it was the ability of attracting other masses, and it was the propensity to be attracted by other masses. Einstein introduced "rest" mass, an aspect of energy, expressed by the equation $E=mc^2$. Einstein also showed that things that possess "mass" cannot travel faster than the speed of light, a speed that is reserved for things that do not possess mass (such as the photon).

Quantum Mechanics showed that "mass" is indeed a property of every elementary particle, but introduced another oddity: while there is an anti-particle for every particle (electrical charge can be positive or negative), both a particle and its anti-particle have the same (positive) mass. Mass is only positive, never negative. In fact, in 1957 British physicist Hermann Bondi showed that the encounter between a mass and its anti-mass would result in infinite acceleration, with no need for a source of energy: the negative mass would be attracted to the positive mass, while the positive mass would be repelled by the negative mass. Thus the two masses would experience equal accelerations in the same direction, in violation of Newton's third law, and continue to accelerate forever, (the negative mass chasing the positive mass and the positive mass fleeing from the negative mass with constant acceleration).

Neither Relativity nor Quantum Theory explained what "mass" is (where it comes from) and what causes its odd properties. They both took it for granted that Nature is that way. It was the odd behavior of "mass" that allowed physicists to create an elegant world. But the elegance was mostly based on an abstract, arbitrary, "catch all" definition.

### *The World And The Mind*

Relativity Theory and Quantum Theory said something important about the mind. They were as much about mind as they were about matter, only in a more subtle way.

Relativity Theory was not only about reality being "relative" to something. It was (first and foremost) about reality being beyond the reach of our senses.

Einstein's underlying principle is that we don't always see the universe as it is. Newton's underlying principle was that we see the universe as it is. Newton's Physics is a description of how our mind perceives the universe. There are bodies, there is absolute time, etc.

Einstein's Physics is a "guess" about what the universe really is, even if our mind cannot perceive it. Einstein's Physics implied that there may be aspects of the universe that our mind cannot perceive, and that we can guess only by analyzing the aspects that we can perceive.

Quantum Theory was not only about reality being "quantized". It was also about reality being beyond the reach of our mind. The single most distressing finding of Quantum Theory is that reality, as we know it, only occurs when somebody observes it. The electron is in a certain place only when somebody actually looks at it, otherwise the electron is, simultaneously, in several different places.

We can analyze this finding with either of two stances. According to the first one, our mind has no limitations. It can perfectly perceive nature as it is. It observes only one value because that is what nature does: the multiple choices for a quantity's value collapse to just one value when that quantity is observed by an observer.

According to the second one, our mind has limitations. The quantum collapse from many values to just one value is due to a limitation of our mind. Our mind cannot perceive nature as it is. It can only perceive one value for each quantity.

The electron is in many places, but our mind cannot perceive a thing being in many places at the same time, so it "collapses" the electron into only one specific place at a time. This is just an effect due to the limitation of our mind. We are forced to "sample" reality because we can't handle all of it. After all, that's what all our senses do. They are bombarded all the time with data from the environment, and they only pick up some of those data. We don't perceive every single detail of what is going on around us, we are forced to be selective. The mind turns out to be a sense that also has limited capacity, although the limitation is of a different kind. Each item of reality (a position, a speed, etc) "has" many values. The reason we observe only one value is that our mind can't handle a universe in which quantities have more than one value.

The conceptual revolution caused by Quantum Theory was somewhat deeper than the one caused by Relativity Theory. Reconciling Newton and Einstein is relatively easy: Newton's theory was not false, it was just a special case of Einstein's theory, the one in which the spacetime is Euclidean. Reconciling Newton and Quantum Theory is, on the other hand, impossible: Newton's theory is just false. It seems to work because we insist on assuming that such things as big objects truly exist.

A theory of the mind that does not take into account Relativity is a legitimate approximation, just like a theory of the Earth that does not take into account Relativity is a legitimate approximation. But no theory of the mind can ignore Quantum Theory.

### *The Power of Constants*

At this point we can note that all the revolutionary and controversial results of these new theories arose from the values of two constants. Quantum Mechanics was a direct consequence of Planck's constant "h": were that constant zero, there would be no uncertainty. Relativity Theory was a direct consequence of the speed of light "c" being constant in all frames of reference: were the speed of light infinite, there would be no time dilatation, nor contraction of length.

These two constants were determined, indirectly, by studying two minor phenomena that were still unsolved at the end of the century: the ether and the black body radiation.

The presence of the ether could not be detected by measuring the speed of light through it; so Einstein assumed that the speed of light is always the same.

The black body does not radiate light with all possible values of energy but only with some values of energy, those that are integer multiples of a certain unit of energy; so Planck assumed that energy exchanges must only occur in discrete packets.

These two universal constants alone revealed a whole new picture of our universe.

They are more than "constants" though: they can also be viewed as powerful operators that dramatically alter reality. By multiplying a time by the speed of light, we obtain a length: the speed of light converts time into space (and vice versa). By multiplying a frequency by the Planck constant, we obtain an energy: the Planck constant converts waves into particles (and vice versa).

Furthermore, the gravitational constant "G" in General Relativity performs a similar function: it converts energy-momentum density into spacetime curvature.

These constants ("h", "c" and "G") acquire a truly metaphysical aspect.

**Quantum Reality: Fuzzy or Incomplete?**

Many conflicting interpretations of Quantum Theory were offered from the beginning.

Niels Bohr claimed that only phenomena (what appears to our senses, whether an object or the measurement of an instrument) are real, in the human sense of the word: particles that cannot be seen belong to a different kind of reality, which, circularly, cannot be perceived by humans; and the wave function is therefore not a real thing. Reality is unknowable because it is inherently indeterminate, and we humans do not live in a world of indeterminate things, we live in a world of phenomena (where "phenomena" presumably includes also houses and trees, the effect of those elementary processes).

Werner Heisenberg, the man who discovered in 1925 the first complete theory of the quantum, believed that the world "is" made of possibility waves and not particles: particles are not real, they are merely "potentialities", something in between ideas and actualities. Our world, what we call "reality", is a sequence of collapses of wave of possibilities. De facto, Heisenberg started a metaphysical revolution by surrendering the concept of reality in favor of the concept of observables: we can't know what really exists, we can only know what we can observe. The quantum wave talks about a new kind of reality, something in between classical possibility and classical reality. The quantum world and our world are bridged by the "measurement". Reality arises from quantum discontinuities (or "quantum jumps"): classical evolution of the Schroedinger equation builds up "propensities", then quantum discontinuities (the collapse of the wave function) select one of those propensities. Every time this happens, reality changes. Therefore reality "is" the sequence of such quantum discontinuities. What turns the unknowable world of particles into human-perceivable "phenomena" is the observation: the moment we observe something, we create a phenomenon. As John Wheeler put it, "no phenomenon is a real phenomenon until it is an observed phenomenon". The universe had to wait for a conscious observer before it started existing for real. Furthermore, Heisenberg interpreted this reality as "knowledge":

the quantum state is a mathematical description of the state of the observer's knowledge rather than a description of the objective state of the physical system observed.

Heisenberg believed in a world of particles, Schroedinger believed in a world of waves. Heisenberg believed that quantum jumps "were" reality, whereas Schroedinger thought he had eliminated them.

The British physicist Paul Dirac, the man who in 1928 merged Quantum Physics and Special Relativity in Quantum Field Theory, pointed out that Quantum Physics is about our knowledge of a system. It does not describe reality but our knowledge of reality. A wave function represents our knowledge of a system before an experiment and the reduced wave function our more precise knowledge after the measurement.

*Entanglement*

Albert Einstein, Boris Podolsky and Nathan Rosen famously argued against Quantum Mechanics describing what they thought was an obvious paradox, which came to be known as the "EPR paradox" ("Can Quantum-Mechanical Description of Physical Reality be Considered Complete?", 1935): according to the equations of Quantum Mechanics, if the states of two particles are ever entangled in a "wave", they will always be, i.e. a change in one of the particles will have an immediate effect on the state of the other particle no matter how far apart they have moved in the interval, thus defying the limit that messages can be broadcast in this universe (the speed of light).

In the next 20 years only about 5 papers would quote the EPR paper. By 2015 that paper would be cited on average more than once a day.

Schroedinger was aware of the issue and immediately responded to Einstein ("Discussion of probability relations between separated systems", 1935) arguing that, on the contrary, "entanglement" (a term coined in that paper) is the most fundamental property of Quantum Mechanics. Schroedinger had good reasons to think so. For example, teleportation sounds impossible not only in the classical world but also in the quantum world, because Heisenberg's uncertainty principle seems to rule out the possibility of making an exact copy of a particle's state (the very act of measuring its state destroys it so it is impossible to make an exact replica), but entanglement offers a way around it. A team led by US physicist William Wootters first reported how to achieve quantum teleportation using entanglement ("Teleporting an unknown quantum state via dual classical and Einstein-Podolsky-Rosen channels", 1993). In 1998 the team led by US physicist Jeff Kimble teleported a photon for about one meter, and in 2003 a Swiss team led by Nicolas Gisin teleported a photon to a distance of 2 kms, and in 2012 a team led by the Austrian physicist Anton Zeilinger achieved a whopping 143 kms).

Entangled quantum states of more than two particles were first studied by the Austrian physicist Anton Zeilinger and the US physicists Daniel Greenberger and Michael Horne ("Bell's theorem without inequalities", 1990).

## *The Indivisible Universe*

Einstein was so unhappy with the uncertainty principle that he accepted Quantum Mechanics only as an incomplete description of the universe. He thought that Quantum Mechanics had neglected some "hidden variables". Once those hidden variables were found, we would have a complete theory without Quantum Theory's oddities but with all of Quantum Theory's results.

Quantum Theory is a practical tool to calculate probabilities for sets of particles, but no prescription is provided for calculating quantities of individual particles. Einstein thought that there is an underlying reality where determinism rules and the behavior of the individual particle can be predicted. It is just that Quantum Mechanics is incomplete and has not found out that underlying reality yet.

Einstein was particularly unhappy about the "nonlocality" of Quantum Physics, which he thought constituted a paradox. "Nonlocality" means "action at a distance". In Quantum Physics one can prove that, if they were once part of the same state, two particles will always be connected: once we measure the position of the first one, we instantaneously determine the position of the other one, even if, in the meantime, it has traveled to the other end of the universe. Since no information can travel faster than light, it is impossible for the second particle to react instantaneously to a measurement that occurs so far from it. The only possible explanation for this "paradox" was, to Einstein, that the second particle must have properties which are not described by Quantum Mechanics.

Einstein thought that Quantum Physics provides a fuzzy picture of a sharp reality, whereas for Bohr it provides a complete picture of a fuzzy reality.

Einstein was proven wrong in 1964 by the Irish physicist John Bell ("On the Problem of Hidden Variables in Quantum Mechanics", published two years later), whose theorem basically ruled out "local hidden variables", precisely the type that Einstein invoked. Bell's conclusion is that, on the contrary, there are objective, non-local connections in the universe. In other words, two particles, once they have interacted, will keep interacting forever (their wave functions get entangled forever). Einstein believed in the law of locality, i.e. that two objects can interact only if they touch each other or if their interaction is mediated by some other object; but Bell proved that the "wave" is enough to provide interaction. Two measurements can be related instantaneously even if they are located in regions too far apart for a light signal to travel between them. Non-locality, or inseparability, is a fact of nature. Objects are not only affected by forces. They are also affected by what happens to other objects.

(More precisely, Bell showed how to test whether a world of properties that are not due to observation and of separated objects is possible. In 1972 John Clauser carried out an actual experiment to perform the test, and its result proved Einstein wrong: either properties are due to observation, or objects are forever connected, or both. Our world cannot possibly have both an observer-independent reality and entanglement-free objects. To be fair to Einstein, Bell assumed that induction is a valid logical method to prove theorems. And, as Nick Herbert has noted, Bell's theorem is metaphysical, not physical, and ultimately relies on the metaphysical assumption that the world behaves in a classical deterministic manner).

This shattered another belief of classical Physics. Newton believed that objects interact through forces that somehow have to travel from one to the other. A cannonball has to travel from the cannon to the walls before the walls explode; and nothing else in the universe is affected. The sun attracts the earth into an orbit, but it doesn't have any effect on the other stars. These are "local" interactions. Einstein added that forces can only travel as fast as light. Therefore, the impact of a force on an object is delayed by the time it takes for the force to reach that object at a speed which cannot exceed the speed of light. "Locality" became a distance: there is only so much in the universe that can exert a force on me, because only so much of the universe can send its force to me during my lifetime. If I live 80 years, an event that occurs more than 80 light-years away from here will never cause any disturbance on my life. Bell proved that this is not the case, because Quantum Theory prescribes the existence of a non-local "force": once two waves have interacted, they are forever entangled.

Note that Heisenberg's "knowledge interpretation" never had a problem with non-locality: obviously, a change in the observer's knowledge does change the observer's knowledge about the entire system, regardless of how "extended" in space the system is. For example, if I observed the two particles at the beginning, when they were in the same place, and noticed that one is black and the other white, and later I observe the white one, I will "know" that the other one is black even if the other one is light-years away from me.

### *Ontological interpretation*

The US physicist David Bohm believed in an "undivided whole" even before John Bell's theorem. His idea was that the whole universe is entangled in one gigantic wave.

One of Quantum Theory's most direct consequences is indeterminism: one cannot know at the same time the value of both the position and the momentum of a particle. One only knows a probability for each of the possible values, and the whole set of probabilities constitute the "wave" associated with the particle. Only when one does observe the particle, does one particular value occur; only then does the wave of probabilities "collapse" to one specific value.

Bohm's "ontological" interpretation of Quantum Theory ("A Suggested Interpretation of the Quantum Theory in Terms of Hidden Variables", 1952) almost resurrected determinism at the quantum level. Bohm's bold assumption was that the quantum "wave" is a real wave, due to a real potential.

Bohm assumed that the wave function does not represent just a set of probabilities: it represents an actual field. A particle is always accompanied by such a field. This field is a real field and acts upon particles the same way a classical potential does.

Bohm resurrected an interpretation of Quantum Theory that de Broglie ("On the role of continuous waves in wave mechanics", 1927) had abandoned, the theory of an ordinary wave guiding an ordinary particle. The Soviet physicists Dmitri Blokhinzhev ("Criticism of the Philosophical Views of the Copenhagen School", 1951) had made a similar proposal.

The beauty of this assumption is that, with the introduction of this additional potential, something momentous happens to the equations of Quantum Mechanics: position and momentum of a particle are no longer incompatible, they can be measured precisely at the same time, and Heisenberg's principle is defeated.

The behavior of the particle in Bohm's theory is determined by the particle's position and momentum, by whatever force is acting on it, and by the quantum potential.

For Bohm, particles do exist and are always accompanied by a field. An electron is neither a particle nor a wave (field), it is a particle plus a wave (that cannot be separated). But Bohm's wave is not Born's wave: Born's wave is only a function of probabilities that helps compute the particle's position, whereas Bohm's wave is a real wave that guides the particle (therefore also referred to as the "pilot-wave").

Everything is both a particle and a wave, and is acted upon by both a classical potential and a quantum potential (the "pilot wave"). Basically, the wave-function provides an additional potential that, once inserted in the traditional Hamiltonian of classical Physics, yields a well-determined trajectory for each particle (but since the initial position cannot be known, we still can't predict the path of a particle, only notice that there exists a well-determined path prescribed by nature).

Bohm had found an interpretation of Quantum Theory in terms of particles with well-defined position and momentum. What Bohm had done with his assumption was, basically, to add some "hidden variables" (the quantum potential) to the equations, precisely what Einstein had suggested to restore determinism in Physics. (Bohm, incidentally, was dismissed equally by Bohr, who did not believe in hidden variables, and by Einstein, who believed in hidden variables).

### *The Pilot-Wave*
To explain the function of the quantum potential, Bohm introduced the notion of "active in-formation" ("information" as in "give form", for example to a particle's movement). A particle is moved by whatever energy it has (for example, because a force is acting on it) but its movement is guided by the "in-formation" in the quantum field (in the "pilot-wave").

In Physics, a potential describes a field in terms of how, at each point in space, the particle located at that point will be affected by that field. In Newton's physics the effect of the classical potential on a particle is proportional to the magnitude of the field.

Bohm thought that his quantum field, in particular, had to reflect whatever is going on in the environment, including the measuring apparatus. Therefore, the quantum potential depends only on the form, and not on the magnitude, of the quantum field. The "strength" of the quantum potential does not depend on the intensity of the wave but only on the form of the wave. Even a very weak quantum potential can affect the particle. Even a very distant event can affect the particle.

The previous interpretations of Quantum Theory were trying to reconcile the traditional, classical concept of "measurement" (somebody who watches a particle through a microscope) with a quantum concept of "system". Bohm dispensed with the classical notion of "measurement": one cannot separate the measuring instrument from the measured quantity, as they interact all the time. It is misleading to call this act "measurement". It is an interaction, just like any other interaction, and, as Heisenberg's principle states, the consequence of this interaction is not a measurement at all.

### *Implicate Order*
The field that Bohm introduced in the equations to fix Heisenberg's indeterminism represents a "sub-quantum" reality.

Bohm's quantum potential does not act within the four-dimensional geometry of spacetime; it acts beyond it. In a sense, it defines a common pool of information, a way to connect everything together, just like dancers can use the music to move together in harmony.

Bohm thought that this field must be fluctuating rapidly and what Quantum Theory observes is merely an average over time (just like Newton's physics reads a value for quantities that are actually due to the Brownian motion of many particles). Quantum physics deals with mean values of an underlying reality just like Newton's physics deals with mean values of thermodynamic quantities.

At this "sub-quantum" level, quantum effects all but disappear: a particle's position and momentum are well-determined. The mystery of the collapse of the wave function, of the discontinuity in the transition from the quantum world to the classical world, occurs only at the quantum level, whereas Bohm believes there is a deeper level at which the apparent discontinuity of the collapse disappears.

After all, the study of "elementary" particles has shown that even elementary particles can be destroyed and created, which means that they are not the ultimate components of the universe, that there must be an underlying reality, or, in Bohm's terms, an underlying "flux". Bohm thought that the basic problem lay in an obsolete notion of "order".

Thus, Bohm distinguished between the "explicate" order (the world of isolated spacetime thing-events that our senses experience) and the "implicate" order (all thing-events are part of a whole, the "holomovement"). The explicate order emerges from the holomovement. The holomovement contains all instances of explicate order as potentialities.

Cartesian order (the "grid" of space-time events) is appropriate for Newtonian physics in which the universe is divided in separate objects, but inadequate for Quantum and Relativity theories to reflect their idiosyncrasies and in particular the undivided wholeness of the universe that Bohm has been focusing on.

Bohm's solution was to contrast the "explicate order" that we perceive and that Physics describes (the Cartesian order of isolated space-time thing-events) with the "implicate order", which is an underlying, hidden layer of relationships. The explicate order is but a manifestation of the implicate order. Space and time, for example, are "forms" of the explicate order that are derived from the implicate order.

The implicate order is similar to the order within a hologram: the implicate order of a hologram gives rise to the explicate order of an image, but the implicate order is not simply a one-to-one representation of the image. In fact, each region of the hologram contains a representation of the entire image. The implicate order and the explicate order are fundamentally different. The main difference is that in the explicate order each point is separate from the others. In the intricate order, the whole universe is "enfolded" in everything, and everything is enfolded in the whole. In the explicate order "things" become (relatively) independent. In the implicate order, all thing-events are part of a whole, the "holomovement". The explicate order emerges from the holomovement. The holomovement contains all instances of explicate order as potentialities.

Bohm suggested that the implicate order could be defined by the quantum potential, the field consisting of an infinite number of pilot waves. The overlapping of the waves generates the explicate order of particles and forces, and ultimately space and time.

Since Bohm's quantum field is affected by all particles (the pilot-wave that guides all particles is affected by all particles), nonlocality is a feature of reality: a particle can depend strongly on distant features of the environment.

Bohm's universe is one indivisible whole.

Everything in the universe is entangled in everything else, and ultimately in the whole. It does not make sense to analyze particles of subsets of the world as independent and separate parts.

### *Beyond Locality*

Einstein's objection did not die there and is still very much alive, if nothing else because, ultimately, it can be read as an objection to the role that the observer plays in Quantum Theory.

The US physicist Alwyn Scott resuscitated Einstein's hypothesis. Scott argued in favor of an interpretation of Quantum Theory as an approximation to a not yet discovered non-linear theory. The new theory must be non-linear because it is the only way to remove Heisenberg's uncertainty principle, which descends from the linearity of Schroedinger's equation.

Again inspired by Einstein, the Australian philosopher Huw Price thinks that backward causation (that future can influence the past), or advanced action, is a legitimate option. Price believes that our theories are time-asymmetric because we are conditioned by folk concepts of causality. Physical theories are built starting with the assumption that the future cannot influence the past, and therefore it is no surprise that they prescribe that the future cannot influence the past. If we remove our preconceptions about causality, then we can redraw Quantum Physics. Then it turns out that Einstein was right with his hypothesis of hidden variables, and that Quantum Physics provides an incomplete description of the universe. A complete Quantum Physics will not assign any critical role to the observer.

### *Transactional Interpretation*

The US physicist John Cramer ("Generalized Absorber Theory and the Einstein-Podolsky-Rosen Paradox", 1980) proposed a way to remove the observer from the formalism of Quantum Mechanics.

The story begins in 1925 when, de facto, Heisenberg made a metaphysical revolution by surrendering the concept of reality in favor of the concept of observables: we can't know what really exists, we can only know what we can observe. Heisenberg's metaphysical move was essential to discovering a theory that turned out to correctly predict observation.

Sometimes one has to surrender conventional beliefs in order to understand the world. Many thinkers, and Bohr in particular, drew the conclusion that reality cannot be known (a` la Kant), not realizing that reality doesn't necessarily have to be "classical" reality: there could exist a different, non-classical, kind of reality. One has to study carefully what the math is saying.

Heisenberg wrote: "The probability wave... was a quantitative version of the old concept of "potentia" in Aristotelian philosophy. It introduced something standing in the middle between the idea of an event and the actual event, a strange kind of physical reality just in the middle between possibility and reality" (Gifford Lectures, 1956).

The story continues with a theory formulated by the US physicists John Wheeler and Richard Feynman for classical electrodynamics ("Interaction with the Absorber as the Mechanism of Radiation", 1945). From a purely mathematical viewpoint, Maxwell's equations of electromagnetism have two possible solutions: a retarded solution and an advanced one. If one reads those equations literally, any electrically charged particle generates both waves that will arrive after the emission as well as other, perfectly symmetrical, waves that will arrive before the emission. Since the latter make no sense in a world driven by the causality principle (that the effect of an action cannot precede the action), the advanced solutions were routinely discarded. However, Wheeler and Feynman found a formally beautiful way to retain them as part of reality. Charged particles were traditionally considered only as radiation emitters. Wheeler and Feynman viewed them as both emitters and absorbers: charged particles emit their radiation half as retarded and half as advanced. The principle of causality is not violated if one retains both the retarded waves that travel from emitter to absorber and the advanced waves that travel from absorber to emitter. In other words, Wheeler and Feynman described a radiation process as a "transaction" in which the emitter of the radiation and the absorber of the radiation exchange waves: the emitter sends a "retarded" wave to the absorber, and simultaneously the absorber sends an "advanced" wave to the emitter. An observer perceives only that a retarded wave has traveled from the emitter to the absorber. Technically, advanced waves have "eigenvalues" of negative energy and frequency, and they propagate in the negative time direction. Advanced waves are basically the time-reversed counterparts of normal (or retarded) waves. Both "advanced" and "retarded" waves are valid orthogonal solutions of the electromagnetic wave equation, but in conventional electrodynamics the advanced solutions are usually ignored as unnatural, because they violate the law of causality, and only "retarded" solutions are retained. Wheeler and Feynman proposed that the time symmetry in the wave equation reflects a property of nature, that both types of waves actually occur.

Paul Davies extended the Wheeler-Feynman model to Quantum Mechanics ("A Quantum Theory of Wheeler-Feynman Electrodynamics", 1970). The problem with this model of particle interaction was that it was difficult to compute.

Cramer interpreted the retarded component of the Wheeler-Feynman as an "offer wave", and the advanced component of the Wheeler-Feynman as a "confirmation wave". Traditionally, only the emission process had been considered real. Cramer gave reality to the absorption process too. Cramer showed that this "transactional" interpretation had one monumental advantage: it derived the Born Rule (that the probability of an event is proportional to the square of the wave function corresponding to that event), something that had historically been an accidental discovery with no rational explanation other than the fact that it works. Furthermore, the measurement process is no longer mysterious: it corresponds with the moment when "all confirmation waves are returned to the emitter from all absorbers capable of responding". Measurements are simply the consequences of transactions, that include both emissions and absorptions.

In other words, Cramer viewed any quantum event as a "handshake" executed through an exchange of advanced and retarded waves. The exchange of a quantum of energy between a present emitter and a future absorber occurs through a Wheeler-Feynman exchange of advanced and retarded waves. The emitter sends an "offer" wave to the absorber (forward in time). The absorber then returns a "confirmation" wave to the emitter (backwards in time). The transaction is then completed with an "handshake" across space-time, which leads to the transfer of energy from emitter to absorber.

Cramer's "transaction" is explicitly non-local because the future is affecting the past.

The US physicist Ruth Kastner remedied some problems of the original Cramer theory and introduced a new interpretation of Cramer's interpretation: Kastner takes the retarded and the advanced waves as being both "real". The key concept, somewhat aligned with David Lewis' "possible worlds" (that all possible worlds are as real as the actual world), is that possibilities exist, although not in spacetime. "Physical" and "actual" are not the same thing: "actual" is a physical entity that is also experienced, but there are physical entities that are never experienced, and therefore never "actual". There is a sort of meta-reality that takes place in what Kastner calls "the pre-spatiotemporal realm". In this realm Cramer's offer and confirmation waves (the results of emissions and absorptions) are not simply mathematical tools but real events. This realm, it turns out, is Hilbert space. Hilbert space is therefore "real" in Kastner's recasting of ontology.

## *The Measurement Problem*

In a 1926 conversation with Heisenberg, Einstein pointed out that measurement is a physical process itself and cannot be abstracted from the physical theory that requires it (as it was assumed in classical Physics). It is the theory that decides what the physicist can observe. A measurement is a lengthy and complex process that leads from the quantity that is measured to the various elements of the measuring apparatus to the observer. One must have a theory of what happens to the "measured" quantity throughout this path. In order to measure a property of a system via a certain apparatus, one must know the correct laws governing the interaction between the system and the apparatus. When a physicist observes a new phenomenon, s/he is forced to use the old theory (because a new theory of the new phenomenon does not exist yet) but this distorts the measurement. Einstein basically warned Heisenberg that Quantum Physics needed a new way of performing measurements that could not be the same as that of classical Physics.

### *The Discontinuity Of Time*
One of Newton's postulates was that "time flows equably".

The biggest problem with Quantum Theory is how the observed world (the world we know, made of well-defined objects) emerges from the quantum world (a world of mere possibilities and uncertainties, thanks to Heisenberg's principle).

The Hungarian mathematician John Von Neumann (the same one who invented the computer) distinguished between processes of the first and second kinds that occur when one is analyzing the evolution of a system with Quantum Theory. First-kind processes occur in isolated systems, on which no measurements can be carried out, and they closely resemble classical, deterministic evolution of a physical system.

Second-kind processes occur when a measurement is carried out and they are non-deterministic (or at least probabilistic): when an observable is measured, the state of the system suddenly jumps to an unpredictable state (or "eigenstate") associated with the measured eigenvalue of the observable. Unlike classical Physics, in which the new state can be determined from the prior state of the system, Quantum Theory can only specify the probabilities of moving into any of the observable's eigenstates. In quantum lingo, a measurement causes a "collapse of the wave function", after which the observable assumes a specific value. A continuous process of the first kind gives rise to a discontinuous process of the second kind.

Isolated systems obey the Schroedinger equation, observed systems obey Heisenberg's quantum jumps. Quantum Theory therefore implies that something turns a process of the first kind into a process of the second kind when it is observed.

The problem is that Quantum Theory does not prescribe or describe when and how this happens. The flow of time is mysteriously altered by measurements: a system evolves in a smooth and deterministic fashion until a measurement is performed, then it jumps more or less randomly into an eigenstate of the measured observable, from where it resumes its smooth evolution until the next measurement. Time seems to behave in an awkwardly capricious way.

As Bohr pointed out, a measurement also introduces irreversibility in nature: collapse cannot be undone. Once we measure a quantity, a discontinuity is introduced, at that point in time, in the evolution of the wave function. If, after a while, we proceed backwards in time, we would reach the same point from the future with a wave function which could collapse into any of the legal ways, only one of which is the one that originated the future we are coming from. It is very unlikely that we would retrace the same past.

Thus there is another "arrow of time" (besides entropy) that explains why time only flows in one direction.

### *The Measurement as Interaction*

According to Quantum Theory, our universe needs both kinds of processes. Von Neumann tried to figure out how they interact and realized that the answer lies in the "measurement" of the system.

Reality seems to proceed on two parallel tracks. The Schroedinger equation determines (in a deterministic manner) the evolution of the state of the system, but that state is a set of possible states each with its own probability of happening. So long as nobody observes the system, the Schroedinger equation predicts future probabilities of the system. Then Heisenberg's principle causes that wave function to "collapse" whenever the system is observed. The collapse causes the system to choose only one of the possible states. Once the observer has observed the system, only a part of the wave survives and evolves according to the Schroedinger equation. At this point the Schroedinger equation can calculate a new set of possible states. And so forth. The two views are both necessary to explain the evolution of the universe. They are not alternative views of the universe. One complements the other.

Note that the observer does more than just observe something: the observer also decides "what" to observe. That decision has an effect on the state of the system, because it forces the system to choose among all the possible states. Nature's role is really only to choose one of those possible states, and Quantum Theory can only presume that this is done randomly.

Von Neumann pointed out that measurement of a system consists in a process of interactions between the instrument and the system, whereby the states of the instrument become dependent on the states of the system. There is a chain of interactions that leads from the system to the observer's consciousness. For example, a part of the instrument is linked to the system, another part of the instrument is linked to the previous part, and so forth until the interaction reaches the observer's eye, then an interaction occurs between the eye and the brain and finally the chain arrives to the observer's consciousness. Eventually, states of the observer's consciousness are made dependent on states of the system, and the observer "knows" what the value of the observable is. Somewhere along this process the collapse has occurred, otherwise the end result of the chain would be that the observer's consciousness would exhibit the same probabilistic behavior of the observable: if the observer reads one specific value on the instrument, it means that the wave of possibilities has collapsed (has chosen just that one specific value) somewhere between the system and the observer's consciousness. At which point? What exactly causes the "collapse"? The instrument? The lens? The electrons inside the instrument? The observer's retina? The observer's nervous system? The observer's consciousness?

What constitutes a valid observer? Does it have to be big? Does it have to be in the brain? Does it have to be conscious? Does it have to be human?

Von Neumann showed mathematically that Quantum Theory is indifferent: it makes no difference to the statistical predictions of Quantum Theory where exactly this happens and what causes it. But humans are curious and would like to find out.

In a sense, Von Neumann was trying to reconcile "objective being" and "subjective knowing". In classical Physics they are one and the same, but in Quantum Physics they are different, and it is not completely clear how subjective knowing relates to objective being.

Later, the Hungarian physicist Eugene Wigner introduced another step in Von Neumann's thought experiment: what if a friend is part of the chain that leads to the observation? If a friend measures the position of a particle and then relates to me the result, for me the wave "collapses" only when she tells me the result of her experiment. But the wave has already collapsed for her when she carried out the measurement. Did the wave collapse also for me at the same time? If not, do our waves collapse to the same value? Or does each of us live in an independent universe?

Von Neumann's interpretation was in turn interpreted as implying that the observer somehow "creates" reality. Copernicus shocked the human race by telling us that we are not at the center of the world; Quantum Physics is telling us that we (our very consciousness) is at the center of the world. We are gods who create our own universe.

### *The Brain as a Measurement Device*

Quantum Theory is really about waves of possibilities. A particle is described by a wave function as being in many possible places at the same time. When the particle is observed, its wave function "collapses" with definite attributes, including the location it occupies, but such attributes cannot be foreseen until they actually collapse. In other words, the observer can only observe a quantum system after having interfered with it.

Von Neumann highlighted an inconsistency in the standard interpretation of Quantum Theory: the objects to be observed are treated as quantum objects (or waves), while the objects that observe (the instruments) are classical objects, with a shape, a position and no wave. The "measurer" is a natural object as much as the "measured", but we grant it immunity from Quantum Theory. Von Neumann objected to dividing the world into two parts that behaved differently. Quantum Theory unequivocally states that everything is a quantum system, no matter how small or big it is. On the other hand, if everything is a quantum system regulated by a wave of possibilities, what makes it collapse? Von Neumann was led again to postulate that something "different" from a quantum system has the power to cause such a collapse, and that something had to be human consciousness. Nothing in the world is real unless perceived by a mind, as the British philosopher Berkeley had argued centuries before Von Neumann.

What if we built an instrument which is smaller than the system to be observed? What would be a quantum system: the smaller or the bigger, the measurer or the measured?

The range of uncertainty of a particle is measured by Max Planck's constant. Because Planck's constant is so small, big objects have a well-defined position and shape and everything. The features of small objects such as particles are instead highly uncertain. Therefore, large objects are granted an immunity from quantum laws that is based only on their size.

Bohr himself ("Can Quantum-Mechanical Description of Physical Reality be Considered Complete?", 1935), aware of this logical inconsistency, interpreted the indeterminacy principle as saying that the measuring apparatus and the observed object constitute one indivisible whole whose parts cannot be analyzed separately.

### *What Creates Reality*

John Wheeler believes that the collapse can be caused by anything that (aware or unaware) makes a "record" of the observation. An observer is anything in Nature that causes the observation to become public and irreversible. An observer could be a crystal.

Roger Penrose, inspired by work done and initiated by the Hungarian physicist Frigyes Karolyhazy ("Gravitation and Quantum Mechanics of Macroscopic Bodies", 1966), invoked gravity to justify that special immunity: in the case of large objects, the space-time curvature affects the system's wave function, causing it to collapse spontaneously into one of the possibilities. Precisely, Penrose believes that different space-time curvatures cannot overlap, because each curvature implies a metric and only one metric can be the metric of the universe at a certain point at a certain time. If two systems engage in some interaction, Nature must choose which metric prevails. Therefore, he concludes, the coupling of a field with a gravitational field of some strength must cause the wave function of the system to collapse. This kind of self-collapse is called "objective" reduction to distinguish it from the traditional reduction of Quantum Theory which is caused by environmental interaction (such as a measurement). Self-collapse occurs to everything, but the mass of the system determines how quickly it occurs: large bodies self-collapse very quickly, elementary particles would not for millions or even billions of years. That is why the collapse of wave functions for elementary particles in practice occurs only when caused by environmental interaction.

In practice, the collapse of the wave, which is the fundamental way in which Quantum Theory can relate to our perceptions, is still a puzzle, a mathematical accident that still has no definite explanation.

It is not clear to anybody whether this "collapse" corresponds to an actual change in the state of the particle, or whether it just represents a change in the observer's amount of knowledge or what. It is not even clear if "observation" is the only operation that can cause the collapse. And whether it has to be "human" (as in "conscious") observation: does a cat collapse the wave of a particle? Does a rock?

What attributes must an object possess to collapse a wave? Is it something that only humans have? If not, what is the smallest object that can collapse a wave? Can another particle collapse the wave of a particle? (In which case the problem wouldn't exist because each particle's wave would be collapsed by the surrounding particles).

What is the measuring apparatus in Quantum Physics? Is it the platform that supports the experiment? Is it the pushing of a button? Is it a lens in the microscope? Is it the light beam that reaches the eye of the observer? Is it the eye of the observer? Is it the visual process in the mind?

It is also a mystery how Nature knows which of the two systems is the measurement system and which one is the measured system: the one that collapses is the measured one, but the two systems are just systems, and it is not clear how Nature can discriminate the measuring one from the measured one and let only the latter collapse.

### *Consciousness Creates Reality*

If a wave collapses (i.e., a particle assumes well-defined attributes) only when observed by a conscious being, then Quantum Theory seems to specify a privileged role for the mind: the mind enters the world through the gap in Heisenberg's uncertainty principle. Indeed, the mind "must" exist for the universe to exist, otherwise nobody would be there to observe it and therefore the world would only consist of possibilities that never turn into actualities. Reality is just the content of consciousness, as the Hungarian physicist Eugene Wigner pointed out ("The Problem of Measurement", 1963). Of course, mind must therefore be an entity that lies outside the realm of Quantum Theory and of Physics in general. The mind must be something special that does not truly belong to "this" world.

Wigner pointed out that to every action there is a reaction: why shouldn't there be a reaction to a conscious observation of a physical phenomenon? If a phenomenon exerts an influence on my consciousness when I observe it, then my consciousness must exert an influence on the phenomenon. Otherwise the fundamental tenet that to every action there is a reaction is no longer true.

Wigner observed that Schroedinger's equation is linear, but would stop being linear if its object were the very consciousness that collapses the wave. Therefore, Schroedinger's equation (which is linear) would result in a non-linear algorithm that may justify the mind's privileged status.

If the collapse occurs only when observed by a conscious being, if the collapse occurs at the border between mind and matter, as Wigner believes, then the evolution of the universe changed after the appearance of human beings (there was no collapse anywhere before mind appeared).

Undeterred by this objection, the US physicist John Wheeler believes that ours is a "participatory" universe, one in which consciousness participates in creating reality. The observer and the phenomenon are engaged in a creative act that yields reality. Consciousness does not create reality. Consciousness's role is extremely limited: it can't even choose which of the possibilities contained in the wave function will become reality. It can only "precipitate" reality out of many possibilities. Which possibility becomes reality is up to nature. Nonetheless, Wigner and Wheeler believe that consciousness is crucial to creating reality: as limited as its contribution is, without it there would be no reality, only possibilities. Wheeler even speculated that the rise of consciousness retroactively determined the history of the universe because it collapsed the mother of all waves that had never been collapsed before, thereby fixing every single event in the previous life of the universe.

Quantum theoretical effects could be considered negligible if they only affected particles. Unfortunately, Erwin Schroedinger, with a famous thought experiment ("The present situation in quantum mechanics", 1935), established that Heisenberg's uncertainty affects big objects too. Basically, Schroedinger devised a situation in which a quantum phenomenon causes the cat to die or stay alive. Since any quantum phenomenon is uncertain, the cat's life is also uncertain: until we look at the cat, the cat is neither alive nor dead, but simply a wave of possibilities itself. (A popular objection to Schroedinger's argument is that the cat can never be in a superimposed state because every big object, by definition, is never isolated, it is always entangled with the rest of its surroundings, and therefore it is "collapsed" all the time).

### *Meta-reality*
As the Israeli physicist Yakir Aharonov and the US philosopher David Albert have pointed out ("Can we make sense out of the measurement process in relativistic quantum mechanics?", 1981), the measurement process of Quantum Theory seems to defy Relativity.

Not only does the "collapse" of the quantum wave create reality as we know it, but it does so instantaneously. Any instantaneous phenomenon contradicts Relativity: according to the principle of Relativity, what is instantaneous for me may not be for you if you are in motion relative to me. This collapse, that is routinely observed in millions of experiments around the world, is empirically real but it is not quite an event: it is not located anywhere in spacetime. It seems to imply the existence of a realm that is "sub-empirical", that provides the foundations for the empirical reality that we observe.

### *Inventing Reality*
Richard Feynman ("Space-Time Approach to Non-Relativistic Quantum Mechanics", 1948) offered yet another interpretation of Quantum Theory: he assumed that all possible states allowed by a wave function exist at any moment. In other words, he took Schroedinger's equation to the letter. The state that is revealed by measurement is merely the state which represents the "path of least action" for the particle relative to the observer. But the particle is in every place allowed by its wave function. An observation does not reveal reality: an observation is an interaction between the observer and the observed system, and the observation simply reveals that: the interaction between the observer and the observed system. In a sense, the observer "invents" the particle. The particle per se does not exist (or, better, it is merely a field). What exists is a range of values, or, better, a set of ranges of values, which our observations translate into values for attributes of a particle.

### *The Multiverse: The Quest for Certainty*
The traditional (or "Copenhagen") interpretation of Quantum Mechanics seems to be trapped in its unwavering faith in uncertainty. Others have looked for ways out of uncertainty.

One possibility is to deny that the wave function collapses at all. Instead of admitting a random choice of one of many possibilities for the future, one can subscribe to all of the possibilities at the same time. In other words, the probabilistic nature of Quantum Mechanics allows the universe to unfold in an infinite number of ways.

Hugh Everett's "many-universes" interpretation of Quantum Mechanics ("Relative State Formulation of Quantum Mechanics", 1957) states, basically, that if something physically can happen, it does: in some universe. Everett interpreted quantum "possibilities" as actualities. A particle "is" in many places at the same time: those places are in different universes. Physical reality consists of a collection of universes: the "multiverse". We exist in one copy for each universe and observe all possible outcomes of a situation. It is not only the universe that splits into many universes, it is also the observer who splits into many observers. For a particle there is no wave of possibilities: each possibility is an actuality in one universe. (Alternatively, one can say that there is one observer for each possible outcome of a measurement).

Each measurement splits the universe in many universes (or, as Michael Lockwood puts it, each measurement splits the observer). Biographies form a branching structure, and one which depends on how often they are observed.

No reduction/collapse occurs. The wave function evolves in a deterministic way, just like in Newton's physics.

Naturally, the observer perceives exactly what I am perceiving: a flow of smooth changes.

There is an alternative way to present Everett's ideas. Everett basically accepts that the Schroedinger equation is all there is. The world is described by that equation. We have to take it literally. The particle is in all the states that the equation prescribes. The trick is that the state of the observer is as superposed as that of the observed system. Therefore the observer sees all of the possible states of the observed system. This way the world does not split, but the mind of the observer does. Each mind observes only one state of the many that are possible according to the Schroedinger equation. Therefore each mind perceives a separate world, that is a subset of the world described by the Schroedinger equation. In a sense, each mind views the world from a subjective perspective. The objective state of the world is the one described by the equation, and it corresponds to the superposition of all the states observed by all the minds of the observer.

The British physicist Stephen Hawking is even trying to write down the wave function of the universe, which will actually describe an infinite set of possible universes. Basically, he looks at the universe as if it were one big particle. Just like the wave function of a particle describes an infinite set of possible particles, the wave function of the universe actually describes an infinite set of possible universes.

In Everett's multiverse, Quantum Theory is deterministic and the role of the observer is vastly reduced (we really don't need an observer anymore, since the wave collapses in every single universe, albeit in different ways).

Quantum Theory looks more like classical theory, except for the multiplication of universes.

### *The Immanent Manyverse*
Because of the apparent approximation of any quantum description of a phenomenon, the Israeli physicist David Deutsch also thinks that our universe cannot possibly constitute the whole of reality, that it has to be part of a "multiverse" of parallel universes. But Deutsch's multiverse is not a mere collection of parallel universes, with a single flow of time. He highlights the contradiction in assuming an external, superior time in which all spacetimes flow. This would still be a classical view of the world. Deutsch's manyverse is instead a collection of moments. There is no such thing as the "flow of time". Each "moment" is a universe of the manyverse. Each moment exists forever, it does not flow from a previous moment to a following one. Time does not flow because time is simply a collection of universes. We exist in multiple versions, in universes called "moments".

A key concept is "fungibility": it means that a set of objects can be considered as a set of identical objects. For example, if I lend you one dollar and a few days later you give me back one dollar, we assume that the state of the world is the same as it was before the borrowing even though the dollar bill that you return to me is not the one that I gave you. However, if I lend you a book and you return me a different book, I would be pretty upset: not all books are alike the way all dollar bills are alike. Photons are fungible: you can't tell one from the other. The atoms of lasers are fungible: they are all the same thing. It turns out that fungible objects can deviate from each other and become different entities... in different universes of the multiverse. And that is the origin of the apparent randomness that an individual in one universe observes. If one views a particle as a multiversal object, randomness and uncertainty disappear: a particle has multiple positions and multiple speeds in multiple universes. The wave associated with a particle is not due to the duality of particles and waves: a particle is distributed across many universes, and therefore it "is" a wave in the multiverse.

Each version of us is indirectly aware of the others because the various universes are linked together by the same physical laws, and causality provides a convenient ordering. But causality is not deterministic in the classical way: it is more like predicting than like causing. If we analyze the pieces of a jigsaw puzzle, we can predict where some of the missing pieces fall. But it would be misleading to say that our analysis of the puzzle "caused" those pieces to be where they are, although it is true that their position is "determined" by the other pieces being where they are.

Furthermore, Deutsch claims that Quantum Theory is not enough to understand reality. He does not adhere to the reductionist stance which says that to understand a system is to understand its parts and to have a theory of that system is to have a set of predictions of its future behavior. Deutsch thinks that the predictions are merely the tools to verify if the theory is correct, but what really matters is the "explanation" that the theory provides. Scientific knowledge consists of explanations, not of facts or of predictions of facts. And, contrary to the dominant "reductionist" approach, an explanation that reduces large-scale events to the movement of the smallest possible constituents of matter is not an explanation. As he puts it, why is a specific atom of copper on the nose of the statue of Churchill? Not because the dynamic equations of the universe predict this and that, and not because of the story of that particle, but because Churchill was a famous person, and famous people are rewarded with statues, and statues are built of bronze, and bronze is made of copper.

Scientists who adhere to the reductionist stance believe that the rules governing elementary particles (the base of the reductionist hierarchy) explain everything but they do not provide the kind of answer that we would call "explanation".

So we need four strands of science to understand reality: a theory of matter (quantum theory), a theory of evolution, a theory of knowledge (epistemology), and a theory of computation. The combined theory provides the "explanations" that Deutsch is interested in.

### *Einselection: Darwinian Collapse*

One man who has been studying the problem of how classical Physics emerges from Quantum Physics (how objects that behave deterministically emerge from particles that behave probabilistically, how coherent states of Quantum Mechanics become classical ones) is the Polish-born Wojciech Zurek. He does not believe that consciousness has anything to do with it: it is rather the environment that determines the emergence of reality.

Since 1991, experiments have been performed to show the progressive evolution of a system from quantum to classical behavior. The goal is to observe the progressive collapse of the wave function, the progressive disappearance of quantum weirdness, and the progressive emergence of reality from probability.

Zurek (1984) has proposed a different twist to the debate on the "collapse of the wave". It doesn't necessarily take an observer. Zurek thinks that the environment destroys quantum "coherence" (superposition). The environment includes anything that may interact with the quantum system, from a single photon to a microscope. The environment causes "decoherence" (the choice of one or some of the possible outcomes) and decoherence causes selection (or "einselection") of which possibilities will become reality. The "best fit" states turn out to be the classical states. Systems collapse to classical states because classical states are the ones that best "fit" the environment.

The environment causes the collapse of the wave just like an observer. Decoherence occurs to any system that interacts with other systems. Large objects are classical and not quantum objects because they are inherently "decohered" by being a collection of interacting parts. Small objects are isolated to some extent and therefore exhibit quantum behavior.

The US physicist James Anglin, a close associate of Zurek, studied the evolution of "open quantum systems" far from equilibrium, which resemble Prigogine's studies on open classical systems ("Decoherence of Quantum Fields", 1997).

This line of research was, indirectly, establishing intriguing similarities between the emergence of classical systems from quantum systems and the emergence of living systems from non-living systems.

*Spontaneous collapse*

In the continuing quest to remove randomness from Quantum Theory, the Italian physicist Giancarlo Ghirardi ("Unified Dynamics For Microscopic And Macroscopic Systems", 1986) proposed yet another way to "complete" Quantum Theory and explain the collapse of the wave function. He argued that the actual measurement on a particle forces it to interact with the measuring equipment. The US physicist Philip Pearle ("Combining stochastic dynamical state-vector reduction with spontaneous localization", 1989) refined that model with "continuous spontaneous localization" due to fluctuations in a ubiquitous field that varies across time and space. Ghirardi's theory, however, was not compatible with Relativity. Roderich Tumulka ("A Relativistic Version of the Ghirardi-Rimini-Weber Model", 2006) devised a way to make it merge with the spacetime geometry of Relativity and to account for John Bell's "entanglement" (an apparent violation of Relativity's postulate that nothing can travel faster than light). The price to pay was that the future state of a system would depend on past states as well as the present state (nonlocality not only in space but also in time). Later the British physicist and investment banker Daniel Bedingham ("Relativistic State Reduction Dynamics", 2010) found another way to harmonize spontaneous localization and Relativity (by introducing a mediating field).

Their ideas are expressed mathematically as a nonlinear stochastic term that is added to the linear Schrödinger equation. The Schrödinger wavefunction then becomes unstable. The bigger the system, the more unstable the equantion. A system made of just a handful of particles is not particularly affected by the nonlinear term, but a system made of many particles is forced to spontaneously collapse immediately. Therefore Schroedinger's cat is both dead and alive for just a split second.

## Qubits

In the 1990s another interpretation of quantum mechanics has been put forth by the Austrian physicist Anton Zeilinger.

He set out to find a fundamental principle that would explain the three odd features of the quantum world:
- quantization (all fundamental physical quantities come in discrete amounts),
- randomness (we can only know the probability of an event) and
- entanglement (everything is connected, no matter how far objects are).

He proposed a very simple principle: each elementary system, called "qubit" (e.g., the spin of the electron), carries one and only one bit of information; two systems carry two and only two bits of information; and so forth.

After all, our physical description of the world is represented by propositions, and each proposition can be true or false, i.e. each elementary system carries one and only one bit of information.

The consequences of this principle are simple to derive:
- I can't know two things about an electron, but only one thing at a time (uncertainty), everything has to be quantized because the unit of information is the bit (yes/no, or one/zero);
- Two systems carry exactly two bits of information, which means that they are entangled forever (if one changes, the other one has to change too in order to still yield two bits of information).
- Schroedinger's equation can be derived as the description of motion in a three-dimensional information space.

Zeilinger's interpretation, therefore, is that only information truly exists and that quantum mechanics is simply a theory of how information behaves in this world.

## Information Loss

The Dutch physicist Gerard't Hooft (Quantum information and information loss in general relativity", 1995) argued that Quantum Physics is Classical Physics after an information loss. This is yet another variation on Einstein's "hidden variables" theory. Hooft noticed that classical variables can take any value, whereas quantum variables can take only some values. Thus, de facto, a classical system gives rise to a quantum system when it loses information. He thinks that this information loss can be due to some "dissipative forces". We know that different starting conditions can lead to the same results because of dissipative forces in the macro-world. For example, if you throw two coins from the top of a skyscraper at different speeds, air friction will cause them to reach the ground at the same speed. The observer will conclude that nature only allows some discrete values for the speed of the coins, when in fact it is air friction that caused them to land at the same speed. Nature is classical at its most fundamental level, but quantum at the level of the laboratory because of "dissipation".

## *Quantum Bayesianism*

The US scientists Christopher Fuchs and Carl Caves and the German scientist Ruediger Schack ("Quantum probabilities as Bayesian probabilities", 2002) tried to explain away the uncertainty of Quantum Mechanics by interpreting the wave function as describing the state of the observer, not of the system. They argued that the wave function's probabilities could as well be interpreted as information about the observer, not about the system; in other words, as "Bayesian" probabilities, a measure of the observer's subjective belief, not of the system's state. In this alternative view, the wave function is interpreted as describing the observer, not the world. Just like Bohr, Fuchs does not consider the wave function to be real. The "collapse" of the wave function represents a change in the observer's beliefs due to having performed an experiment. This also means that the same system has as many wave functions as there are observers. The catch, of course, is that in this interpretation we know virtually nothing about the world: the equations of Physics only tell us something about the observers, i.e. about us.

## *Interpretations of Classical Physics*

The fact that Quantum Physics lends itself to many contradicting interpretations has been widely publicized from the very beginning. Less publicized is the fact that Newton's Physics is no less open to interpretations.

Newton's greatest invention was the concept of "mass". Ask ten scientists what "mass" is and you will get ten different answers. Mass is at least three things in Newton's Physics: a measure of resistance to acceleration, a measure of how much an object attracts other objects, and a measure of how much an object is attracted by other objects (laziness, allure and weakness). Whichever of the three you choose, where does it come from? Why do objects have this exoteric quantity of "mass"?

Another fundamental tenet of Newton's Physics (that actually comes from Galileo) is the notion that objects tend to move in a straight line at constant speed. Aristotle thought that objects tend to stop if they are not pushed. Galileo realized that objects (such as arrows) keep moving even when no force is pushing them. Thus it made sense to assume that objects want to keep moving indefinitely. (Friction and gravity cause them to slow down or bend). This works. But: why do objects have a preference for traveling in a straight line at constant speed? Where does this property come from? Again, this is open to interpretation.

In conclusion, it is not surprising at all that there are several different interpretations of what Quantum Physics means: there are still, three centuries later, different interpretations of what Newton's Physics means.

### *The Interpretation of the Human Brain*
Over its first century of existence, Quantum Physics has been the subject of countless "interpretations". Its implications (that reality is created by the observer, that everything is connected all the time, that the universe is run by randomness) sound "odd" and therefore require that someone "interprets" them for the human mind.

However, one could look at the problem from the opposite viewpoint: what is it in the human brain that makes Quantum Physics look so odd? Maybe there is nothing to interpret in Quantum Physics, but there is something to interpret in the human brain. Maybe another brain would not find Quantum Physics so odd, and it would in fact see the world exactly the way Quantum Physics presents it (with objects in multiple positions at the same time and with everything in the universe connected all the time). The human brain has a "cognitive closure". Just like any other brain, there are things that our brain just cannot do. We cannot see or hear frequencies that other animals can see and hear. There is probably an infinite number of things that our brain just cannot do, because it was not designed for them. Maybe understanding Quantum Physics is one of the many things that our brain just cannot do.

### *The Physics of Elementary Particles: Close Encounters with Matter*
Quantum Theory redrew our picture of nature and started a race to discover the ultimate constituents of matter. This program culminated in the formulation of theories of Quantum Electrodynamics, virtually invented by the British physicist Paul Dirac in 1928 when he published his equation for the electron in an electromagnetic field (which combined Quantum Mechanics and Special Relativity) and Quantum Chromodynamics, virtually invented by the US physicist Murray Gell-Mann when he hypothesized the breakdown of the nucleus into quarks ("A Schematic Model of Baryons and Mesons", 1964).

It follows from Dirac's equation that for every particle there is a corresponding anti-particle which has the same mass and opposite electric charge, and, generally speaking, behaves like the particle moving backwards in space and time.

Forces are mediated by discrete packets of energy, commonly represented as virtual particles or "quanta". The quantum of the electromagnetic field (e.g., of light) is the photon: any electromagnetic phenomenon involves the exchange of a number of photons between the particles taking part in it. Photons exchange energy in units of the Planck constant, a very small value, but nonetheless a discrete value.

Other forces are defined by other quanta: the weak force by the W particle, gravitation by the graviton and the nuclear force by gluons.

Particles can, first of all, be divided according to a principle first formulated (in 1925) by the Austrian physicist Wolfgang Pauli: some particles (the "fermions", named after the Italian physicist Enrico Fermi) never occupy the same state at the same time, whereas other particles (the "bosons", named after the Indian physicist Satyendra Bose who made the discovery in 1924) do. The wave functions of two fermions can never completely overlap, whereas the wave functions of two bosons can completely overlap (the bosons basically lose their identity and become one).

(Technically, "boson" is the general name for any particle with an angular momentum, or spin, of an integer number, whereas "fermion" is the general name for any particle with a odd half quantum unit of spin).

It turns out (not too surprisingly) that fermions (such as electrons, protons, neutrons) make up the matter of the universe, while bosons (photons, gravitons, gluons) are the virtual particles that glue the fermions together. Bosons therefore represent the forces that act on fermions. They are the quanta of interaction. An interaction is always implemented via the exchange of bosons between fermions.

(There exist particles that are bosons but do not represent interactions, the so called "mesons", first hypothesized by the Japanese physicist Hideki Yukawa in 1935. Mesons decay very rapidly. No stable meson is known).

Three forces that act on elementary particles have been identified: the electromagnetic, the "weak" and the "strong" forces. Correspondingly, there are bosons that are weak (W and Z particles), strong (the gluons) and electromagnetic (the photon).

Fermions can be classified in several ways. First of all, the neutron and the proton (the particles that made up the nuclei of atoms) are not elementary: they are made of 18 quarks (six quarks, each one coming in three "colors"). Then there are twelve leptons: the electron, the muon, the tau, their three neutrinos and their six anti-particles. A better way to organize Fermions is to divide them in six families, each led by two leptons: the electron goes with the electron's neutrino, the down quark and the up quark. This family makes up most of the matter we know. Another family of Fermions is led by the muon and contains its neutrino and contains two more quarks. The third family contains the tau particle, its neutrino and two more quarks (bottom and top).

Particles made of quarks are called "hadrons" and comprise "baryons" (made of three quarks, and therefore fermions, such as the proton and the neutron) and "mesons" (made of one quark and one antiquark, and therefore bosons). These particles are kept together by the strong forced mediated by gluons.

The electromagnetic force between leptons is generated by the virtual exchange of mass-less particles called "photons". The weak force is due to the W and Z particles (there are two W particles). The "strong" force between quarks (the one that creates protons and neutrons) is generated by the virtual exchange of "gluons". Quarks come in six "flavors" and three "colors". Gluons are sensitive to color, not to flavor. The "strong" force between protons and neutrons is a direct consequence of the color force mediated by gluons. Note that gluons, unlike photons, also interact among themselves because they carry and react to color charges (photons don't carry and don't react to the electric charge).

Leptons do not have color, but have flavor (for example, the electron and its neutrino have different flavors). The "weak" force is actually the flavor force between leptons. W+ and W- are the quanta of this flavor force.

Before the discovery of quarks, we knew of no force that grows with distance. Quarks don't exist as individual particles because they can never be separated: the stronger you pull them apart, the stronger their interaction.

From the point of view of Quantum Field Theory, quantum fluctuations cause vibrations that we perceive as attributes of particles: mass, spin, charge, etc. Thus a particle (e.g., an electron) can be viewed as a vibrational pattern (a pattern of vibrations in a quantum field). Vibrations interact among themselves. When we observe this interaction, we perceive it as particles interacting with each other by exchanging virtual particles (that we interpret as forces, e.g. electromagnetism).

This model explains what we know of matter. It does not explain why there are four forces, 6x3 quarks, six leptons, four bosons for leptons, and eight gluons for quarks. The numbers seem to be arbitrary.

In particular, it does not explain why particles have the masses they have; rather arbitrary numbers that span ten orders of magnitude, the top quark being about 100,000 times "heavier" than the electron, and more than one trillion times heavier than the electron-neutrino. A field called the Higgs field, mediated by the Higgs boson (an idea proposed in 1964 by the British physicist Peter Higgs and confirmed experimentally in 2012) is supposed to permeate the universe, and the mass of a particle is supposed to be a measure of the intensity of its interaction with the Higgs field. Mass would therefore be an emergent property rather than a fundamental property of matter. The Higgs boson would be the only particle equipped with mass from the beginning. The US physicist William Bardeen showed that the mass of the Higgs boson could be a consequence of scale-symmetry breaking ("On naturalness in the standard model", 1995) and scale-symmetry breaking could also explain the exponential inflation that followed the Big Bang, argued by the Italian physicists Alberto Salvio and Alessandro Strumia ("Agravity", 2014). Bardeen's idea is that scale (size) would not be an inherent feature of nature (galaxies would not necessarily be bigger than stones) but would emerge spontaneously from the interaction of the particles constituting the universe.

Nor does the theory explain why this universe needs three fundamental forces (weak, strong and electromagnetic). Roughly, the strong force (that acts only between quarks) creates nuclei. The electromagnetic force (that acts only between charged particles) creates atoms (and whatever is left of it creates molecules). Together these two forces create the matter that we see. The weak force is responsible for the decay of quarks and leptons into different kinds of quarks and leptons. When a quark or lepton decays, it is said to change "flavor". While it is difficult to visualize the weak interaction, it is quite important for our being here because it transforms elements into other elements that account for stars and life (the thermonuclear fusion that transforms hydrogen into deuterium and fuels the Sun is a weak-interaction process, and the life forms of the Earth were made possible by the production of carbon and heavier elements inside stars). Basically, the strong force holds particles together inside the nucleus of the atom, whereas the weak force changes the nature of those particles.

Alternatively, leptons and quarks can also be combined in three families of fermions: one comprising the electron, its neutrino and two flavors of quarks ("up" and "down"); one comprising the muon, its neutrino and two flavors of quarks ("strange" and "charmed"); and one comprising the tau lepton ("tauon"), its neutrino and two flavors of quarks ("bottom" and "top"). Plus the three corresponding families of anti-particles. Eight particles per family (each flavor of quark counts as three particles). The grand total is 48 fermions. The bosons are twelve: eight gluons, the photon and the three bosons for the weak interaction. Sixty particles overall.

The profusion of particles is simply comic. Quantum Mechanics has always led to this consequence: in order to explain matter, a multitude of hitherto unknown entities is first postulated and then "observed" (actually, verified consistent with the theory). More and more entities are necessary to explain all phenomena that occur in the laboratory. When the theory becomes a self-parody, a new scheme is proposed whereby those entities can be decomposed in smaller units. So physicists are already, silently, seeking evidence that leptons and quarks are not really elementary, but made of a smaller number of particles. It is easy to predict that they will eventually break the quark and the electron, and start all over again.

Several other characteristics look bizarre. For example, the three families of fermions are very similar: what need did Nature have to create three almost identical families of particles?

The mass of the muon is 209 times the mass of the electron, and the mass of the tau lepton is 3478 times the mass of the electron. Why?

The spins of these particles is totally arbitrary. Fermions have spin 1/2 and bosons have integral spin. Why?

The whole set of equations for these particles has 19 arbitrary constants. Why?

Gluons are fundamentally different from photons: photons are intermediaries of the electromagnetic force but do not themselves carry an electric charge, whereas gluons are intermediaries of the color force that do carry themselves a color (and therefore interact among themselves). Why?

Also, because color comes in three varieties, there are many gluons, while there is only one photon. As a result, the color force behaves in a fundamentally different way from the electromagnetic force. In particular, it extends to infinite. That confines quarks inside protons and neutrons. Why?

### *The Language of Elementary Particles*

Mass, spin, charge and so forth are the fundamental properties of matter. But "what" exactly are they? They seem to be just verbal metaphors to express the fact that there are ways and rules in which particles interact. For example, an electron and a proton interact in a way that can be visualized by presupposing a force called "electrical force" that operates on negative and positive charges and assigning a negative charge to the electron and a positive one to the proton. These properties, ultimately, are about patterns of behavior. If we saw Mr. A buying the newspaper every day from the store of Mr. B, we could conclude that there exists an AB force that operates between negatively charged people and positively charged people that leads the negatively charged ones to buy newspapers from positively charged people and that Mr. A is negative and Mr. B is positive.

### *Of Symmetry and Asymmetry: Parity*

The German physicist Otto Laporte first classified the wavefunctions as either even or odd, according to their symmetry and t discovered what appeared to be a principle of conservation of parity ("The structure of the iron spectrum", 1924): at the end of an atomic interaction the total parity of the involved wavefunctions is zero. If one of the involved wavefunctions switched from odd to even, then another one must have switched in the other direction. The Hungarian physicist Eugene Wigner showed that Laporte's rule was a consequence of the mirror image symmetry of the Schroedinger equation ("Some consequences of Schroedinger's theory for the term structures", 1927) and generalized the law of conservation to nuclear interactions. It felt reasonable to believe that the mirror image of a physical process should be identical in every way to the original one.

Throughout the years there emerged doubts about the conservation of parity peaking with the work of Chinese-born physicists Chen Ning Yang and Tsung-Dao Lee ("Question of Parity Conservation in Weak Interactions", 1956) that led to the experimental confirmation by Chien-Shiung Wu.

## *Of Symmetry and Asymmetry: Gauge Theory*

When James Maxwell turned electricity and magnetism into "field theories", he changed our perception of nature: not a set of observable particles that interact by touching each other, but a landscape of invisible fields that potentially extend to infinite. Energy, in particular, resides in the field: not only in the electrified and magnetic bodies but also in the space surrounding them.

One fact that became apparent is that the same reality (in terms of quantities that we measure such as speed and electrical charge) can be implemented by different fields. A "gauge" transformation is any function that turns one field into another field whose observable quantities are the same, and the phenomenon is called "gauge invariance" or "gauge symmetry". If you can add a constant to one of the variables describing the field, and nothing changes in the quantities that you measure, then you are in the presence of gauge invariance.

The concept of the gauge field was introduced by the German mathematician Hermann Weyl ("Gravitation and Electricity", 1918) while trying (unsuccessfully) to unify gravitation and electromagnetism. A decade later he succeeded in showing that Maxwell's theory in quantum mechanics is invariant (or "symmetric") under a gauge transformation ("Electron and Gravitation", 1929). He realized that one could derive Maxwell's electromagnetism purely from his old gauge principle. Gauge invariance constitutes a symmetry principle and that principle yields Maxwell's electromagnetism. (Incidentally, Weyl's "two-component spinor formalism" introduced in the same paper led him to foresee that conservation of parity, just formulated in 1924 by Otto Laporte, could be violated).

Gauge fields were not used until Chen Ning Yang and Robert Mills ("Conservation of Isotopic Spin and Isotopic Gauge Invariance", 1954) generalized Weyl's electromagnetic gauge principle to the case of non-Abelian algebra, a particular case of Lie algebra, which eventually led to the unification of three of the fundamental forces. Maxwell's equations assume that there is only one kind of charge (the electric one), whereas the Yang-Mills equations allow for many. As Freeman Dyson wrote ("Unfashionable pursuits", 1983): "Quantum chromodynamics… is conceptually little more than a synthesis of Lie's group-algebras with Weyl's gauge fields." Weyl himself had shown the fundamental role played by Lie algebra in quantum mechanics ("Quantum Mechanics and Group Theory", 1927).

The success of the Yang-Mills gauge theory led to the assumption that all fundamental forces are direct consequences of the properties of gauge symmetries.

### *Unification: In Search of Symmetry*

Since the electric charge also varies with flavor, it can be considered a flavor force as well. Along these lines, Steven Weinberg and Abdus Salam ("A Model of Leptons", 1967) unified the weak and the electromagnetic forces into one flavor force, and discovered a third flavor force, mediated by the Z quanta. The unified flavor force therefore admits four quanta: the photon, the W- boson, the W+ boson and the Z boson. These quanta behave like the duals of gluons: they are sensitive to flavor, not to color. All quanta are described by the so called "Yang-Mills field", which is a generalization of the Maxwell field (Maxwell's theory becomes a particular case of Quantum Flavor Dynamics: Quantum Electrodynamics).

The symmetry of the electroweak force (whereby the photon and the bosons get transformed among themselves) is not exact as in the case of Relativity (where time and space coordinates transform each other): the photon is mass-less, whereas bosons have mass. Only at extremely high temperatures the symmetry is exact. At lower temperatures a spontaneous breakdown of symmetry occurs.

This seems to be a general caprice of nature. At different temperatures symmetry breaks down: ferromagnetism, isotropic liquids, the electroweak force... A change in temperature can create new properties for matter: it creates magnetism for metals, it creates orientation for a crystal, it creates masses for bosons.

The fundamental forces exhibit striking similarities when their bosons are mass-less. The three families of particles, in particular, acquire identical properties. This led scientists to believe that the "natural" way of being for bosons in a remote past was mass-less. How did they acquire the mass we observe today in our world? And why do they all have different masses? The Higgs mechanism gives fermions and bosons a mass. Naturally it requires bosons of its own, the Higgs bosons (particles of spin 0).

Each interaction exhibits a form of symmetry, but unfortunately they are all different, as exemplified by the fact that quarks cannot turn into leptons. In the case of the weak force, particles (e.g., the electron and its neutrino) can be interchanged, while leaving the overall equations unchanged, according to a transformation called SU(2), meaning that one particle can be exchanged for another one. For the strong force (i.e., the quarks) the symmetrical transformation is SU(3), meaning that three particles can be shuffled around. For the electromagnetic force, it is U(1), meaning that only the electrical and magnetic component of the field can be exchanged for each other. Any attempt to find a symmetry of a higher order results into the creation of new particles. SU(5), for example (proposed by Howard Georgi and Sheldon Glashow in 1974), entails the existence of 24 bosons... but it does allow quarks and leptons to mutate into each other (five at the time), albeit at terribly high temperatures.

### *The Quantum is not Relative*

Finally, Quantum Theory does not incorporate gravity. Since gravity is an interaction (albeit only visible among large bodies), it does require its own quantum of interaction, the so called "graviton" (a boson of spin 2). Once gravity is quantized, one can compute the probability of a particle interacting with the gravitational field: the result is... infinite.

The difficulty of quantizing gravity is due to its self-referential (i.e., non-linear) nature: gravity alters the geometry of space and time, and that alteration, in turn, affects the behavior of gravity.

The fundamental differences between Quantum Theory and General Relativity can also be seen topologically: the universe of Relativity is curved and continuous; the universe of Quantum Theory is flat and granular. Relativity prescribes that matter warps the continuum of spacetime, which in turns affects the motion of matter. Quantum Theory prescribes that matter interacts via quanta of energy in a flat spacetime. (Even finding a common vocabulary is difficult!) The bridge between the two views would be to "quantize" spacetime, the relativistic intermediary between matter and matter: then the two formulations would be identical. If spacetime warping could be expressed in terms of quanta of energy, then the two prescriptions would be the same.

### *Superstring Theory: Higher Dimensions*

Countless approaches have been proposed to integrate the quantum and the (general) relativistic views of the world.

The two theories are obviously very different and the excuse that they operate at different "granularity" levels of nature (Quantum Theory for the very small and Relativity Theory for the very big) is not very credible.

Physicists have been looking for a theory that explains both, a theory of which both would be special cases. Unfortunately, applying Quantum Theory to Relativity Theory has proved unrealistic.

The problem is that they are founded on different "metaphors" of the world. Relativity Theory binds together space-time and matter. Quantum Theory binds together matter and the observer (an observer who is supposed to verify the consequences of binding together matter and the observer who is supposed to...).

Relativity focuses on how the gravity of massive bodies bends the structure of time and space and are in turn influenced in their motion by the curvature of space-time. Quantum Theory focuses on the fuzziness in the life of elementary particles.

If one simply feeds Schroedinger's equation (how the world evolves according to Quantum Theory) into Einstein's equation (how the world evolves according to Relativity Theory) the resulting equation appears to be meaningless.

Basically, we don't have a Physics that holds in places where both gravity and quantum effects are crucial, like at the centers of black holes or during the first moments of the "Big Bang".

General Relativity explains motion. Einstein's equations are precise. Quantum Theory explains that motion is undefined. Heisenberg's principle is fuzzy.

General Relativity shows that time is relative. Quantum Theory assumes a universal watch setting the pace for the universe. "Time" looks completely different in one theory and in the other, almost as if the two theories used the term "time" to refer to two different things.

Ditto for the "observer": Einstein's observer is part of the universe and in fact is affected by the universe, whereas Quantum Theory's observer has a special status that exempts her from quantum laws (the quantum universe is divided into particles that are measured and "observers" who make measurements).

## *Superstrings*

A route to merging Quantum Theory and Relativity Theory is to start with Relativity and see if Quantum Theory can be found as a special case of Einstein's equations.

In 1919 the German physicist Theodor Kaluza ("On the Unification Problem of Physics", published only in 1921) discovered that electromagnetism arises if a fifth dimension is added to Einstein's four-dimensional spacetime continuum: by re-writing Einstein's field equations in five dimensions, Kaluza obtained a theory that contained both Einstein's General Relativity (i.e., the theory of gravitation) and Maxwell's theory of electromagnetism. Note that Einstein's Relativity does not say anything about the number of dimensions of our world: it works in any dimensions, unlike Newton's theory that yields the correct (inverse square) formula for the force of gravity only in the case of three spatial dimensions. Kaluza thought that light's privileged status came from the fact that light is a curling of the fourth spatial dimension.

Basically, it was an extension of Einstein's fundamental intuition: gravitation is due to the geometry of a four-dimensional space-time. Kaluza realized that one only has to add a fifth dimension in order to obtain the same statement for electromagnetism: electromagnetism is due to the geometry of a five-dimensional space-time. A theory of gravitation in five dimensions yields both a theory of gravitation and a theory of electromagnetism in four dimensions.

The price to pay is that the fifth dimension behaves in a weird way. The Swedish mathematician Oskar Klein ("Quantum theory and five-dimensional theory of relativity", 1926) explained how the fifth dimension might be curled up in a loop the size of the Planck length (the shortest length that Quantum Physics can deal with). The universe may have five dimensions, except that one is not infinite but closed in on itself. In the 1960s the US physicist Bryce DeWitt ("Quantum Theory of Gravity", 1967) and others proved that a Kaluza theory in higher dimensions is even more intriguing: when the fifth and higher dimensions are curled up, the theory yields the Yang-Mills fields required by Quantum Mechanics.

The Kaluza-Klein theory made a fundamental assumption: the physical world as we know it, and in particular its fundamental forces, originate from the geometry of hidden dimensions. The fundamental forces appear to be "forces" only in a four-dimensional subset of the universe. They are actually just geometry.

It was this approach that in 1970s led the US physicist John Schwarz to formulate Superstring Theory. His early studies had been triggered by a formula discovered by the Italian physicist Gabriel Veneziano ("Construction Of A Crossing-Symmetric, Reggeon Behaved Amplitude For Linearly Rising Trajectories", 1968) and its interpretation as a vibrating string by the Japanese physicist Yoichiro Nambu ("Quark Model And The Factorization Of The Veneziano Amplitude", 1970). Schwarz realized that both the standard model for elementary particles and General Relativity's theory of gravitation were implied by Superstring Theory ("Dual Models For Nonhadrons", 1974).

Superstring Theory views particles as one-dimensional entities (or "strings") rather than points: tiny loops of the magnitude of the Planck length. Particles are simply "resonances" (or modes of vibrations) of tiny strings. In other words, all there is to matter are vibrating strings and each particle is due to a particular mode of vibration of the string. Each vibrational mode has a fixed energy, which means a mass, charge and so forth. Thus the illusion of a particle. All matter consists of these tiny vibrating strings. The key point is that one of these vibrational modes is the "graviton", the particle that accounts for gravitation: Superstring theory is a Quantum Theory that predicts the existence of General Relativity's gravitation.

The behavior of our universe is largely defined by three universal constants: the speed of light, the Planck constant and the gravitational constant. The "Planck mass" is a combination of those three magic numbers and is the mass (or energy) at which the superstring effects would be visible. Unfortunately, this is much higher than the mass of any of the known particles. Such energies were available only in the early stages of the universe and for a fraction of a second. The particles that have been observed in the laboratory are only those that require small energies. A full appreciation of Superstring Theory would require enormous energies. Basically, Superstring Theory is the first scientific theory that states the practical impossibility of being verified experimentally (at least during the lifetime of its inventors).

Furthermore, the superstring equations yield many approximate solutions, each one providing a list of mass-less particles. This can be interpreted as allowing a number of different universes: ours is one particular solution, and that solution will yield the particles we are accustomed with. Even the number of dimensions would be an effect of that particular solution.

There is, potentially, an infinite number of particles. Before the symmetry breaks down, each fermion has its own boson, which has exactly the same mass. So a "photino" is postulated for a "photon" and an "s-electron" for the electron.

Space-time must have ten dimensions. Six of them are curved in minuscule tubes that are negligible for most uses. Matter originated when those six dimensions of space collapsed into superstrings. Ultimately, elementary particles are "compactified" hyper-dimensional space.

Einstein's dream was to explain matter-energy the same way he explained gravity: as fluctuations in the geometry of space-time. The "heterotic" variation of Superstring Theory, advanced by the US physicist David Gross ("Heterotic String", 1985) and others, does just that: particles emerge from geometry, just like gravity and the other forces of nature. The heterotic string is a closed string that vibrates (at the same time) clockwise in a ten-dimensional space and counterclockwise in a 26-dimensional space (16 dimensions of which are compactified).

Some believed that Einstein's General Theory of Relativity is implied by Superstring Theory, to the point that another US physicist, Edward Witten, wrote that Relativity Theory was discovered first by mere accident. Incidentally, the same Witten, provided the most complete "field string theory" yet ("Noncommutative Geometry and String Field Theory", 1986).

In the meantime Superstring Theory progressed towards a peculiar form of duality. A Finnish and a British physicist, Claus Montonen and David Olive ("Magnetic monopoles as gauge particles?", 1977), proposed that there may exist a dual Physics which deals with "solitons" instead of "particles". In that Physics, magnetic monopoles are the elementary units, and particles emerge as solitons, knots in fields that cannot be smoothed out (in our conventional Physics, magnetic monopoles are solitons of particles). Each particle corresponds to a soliton, and viceversa. They proved that it would not matter which Physics one chooses to follow: all results would automatically apply to the dual one.

In particular, one could think of solitons are aggregates of quarks (as originally done in 1974 by the Dutch physicist Gerard't Hooft). Then a theory of solitons could be built on top of a theory of quarks, or a theory of quarks could be built on top of a theory of solitons.

The US physicist Andrew Strominger ("Microscopic origin of the Bekenstein-Hawking entropy ", 1996) found a connection between black holes and strings: if the original mass of the black hole was made of strings, the Hawking radiation (see later) would ultimately drain the black hole and leave a "thing" of zero size, i.e. a particle. Since a particle is ultimately a string, the cycle could theoretically resume: black holes decaying into strings and strings decaying into black holes.

Superstring Theory is the only scientific theory of all times that requires the universe to have a specific number of dimensions: but why ten?

Physicists like the Romanian-born Peter Freund ("Higher Dimensions, Supersymmetry, Strings", 1985) and Michio Kaku observed that the laws of nature become simpler in higher dimensions. The perceptual system of humans can only grasp three dimensions, but at that level the world looks terribly complicated. The moment we move to a fourth dimension, we can unify phenomena that looked very different. As we keep moving up to higher and higher dimensions, we can unify more and more theories. This is precisely how Einstein unified Mechanics and Electromagnetism (by introducing a fourth dimension), how quantum scientists unified electromagnetism with the weak and strong nuclear forces and how particle physicists are now trying to unify these forces with gravity.

Still: why ten?

Are there more phenomena around that we still have to discover and that, once unified with the existing scientific theories, will yield even more dimensions? Are these dimensions just artifices of the Mathematics that has been employed in the calculations, or are they real dimensions that may have been accessible in older times?

One important implication of superstring theory is that the "constants" of nature are fields, and can therefore change in time and in space. The "dilaton" is the field of all fields, that determines the strength of all interactions.

### *Branes*

Another important implication is that our universe might not be all that there is. The electron uses only three of the available dimensions (the so called "Dirichlet membrane" or "D-brane"), and our universe may simply be confined to those three dimensions, but the other dimensions might be "filled" with something else.

D-branes were discovered by the US physicist Joseph Polchinski ("New Connections Between String Theories", 1989) and the Czech physicist Petr Horava ("Strings on world-sheet orbifolds", 1989). They are mathematical objects that resemble membranes in a five-dimensional spacetime. Our universe could in fact be a D-brane extending over the three familiar spatial dimensions. The objects of our world could be stuck to this D-brane like bugs on flypaper. Hence the illusion of three dimensions.

If elementary particles are indeed the different modes of vibration of strings, Polchinski ("Dirichlet Branes and Ramond-Ramond Charges", 1995) realized that these strings are attached to "D-branes", a finding that triggered the so called "Second Superstring Revolution" of M-theory and Holographic Theory.

There might be countless branes around. Each brane contains its own particles, including bosons (i.e., its own portfolio of forces). Ordinary particles are vibrational modes of open strings, strings that are confined to a brane. The particles of a brane are most likely insensitive to the forces that prevail in some other brane. For example, the particles of our brane are sensitive to electromagnetism but not to the many other kinds of force that may exist in many other branes; and, viceversa, the particles of other branes are most likely insensitive to electromagnetism. The higher dimensional space is a sort of mega-brane, containing its own particles and forces. The graviton is the vibrational mode of a closed string, not an open one. It is not confined to branes. It is one of those particles that exist in the higher dimensional space and can therefore communicate with the particles confined into lower-dimension branes. Gravity is a force (the only force?) that is not confined to a brane. The reason that gravity appears to us (inside our brane) so intrinsically different from the other forces is that... it is.

When it comes to our world, we live in a three-dimensional space in which the Standard Model rules. Gravity lives in the higher-dimensional brane. Gravity can travel to our membrane, but it arrives considerably weakened, and that is why it is much weaker than the other forces. The larger the extra dimensions are, the weaker the force that originates from them when perceived inside a brane.

### *M-theory*

The problem with superstring theory is that, quite simply, there are too many superstring theories (at least five), each of them perfectly valid. Edward Witten noted that a few of them were simply the same theory viewed from different angles, and proposed the M-theory, the theory of all possible superstring theories ("Heterotic and type I string dynamics from eleven dimensions", 1995). It turns out that M-theory can also be viewed as an eleven-dimensional theory of ten-dimensional theories.

M-theory does not constrain the universe to be the one we observe: a staggering 1 to the $500^{th}$ power different kinds of universe are technically possible, each with its own laws of nature. The US physicist Leonard Susskind has suggested that all those universes might indeed exist: anything that can exist, does exist. Our universe is merely one item in the "megaverse."

### *Quantum Gravity*

Roger Penrose is one of those who argues that the right approach to the integration of Quantum Theory and Relativity Theory is not to be concerned about the effects of the former on the latter but viceversa.

Penrose (like everyone else) is puzzled by the two different, and incompatible, quantum interpretations of the world. One is due to Schroedinger's equation, which describes how a wave function evolves in time. This interpretation is deterministic and provides a continuous history of the world. The other is due to the collapse of the wave function in the face of a measurement, which entails determining probabilities of possible outcomes from the squared moduli of amplitudes in the wave function ("state-vector reduction"). This interpretation is probabilistic and provides a discontinuous history of the world, because the system suddenly jumps into a new state. We can use Schroedinger's equation to determine what is happening at any point in time; but, the moment we try to actually measure a quantity, we must resort to state-vector reduction in order to know what has happened.

Penrose postulates that these two incompatible views must be reconciled at a higher level of abstraction by a new theory, and such a theory must be based on Relativity Theory. Such a theory, which he calls "quantum gravity", would also rid Physics of the numerous infinites that plague it today. It should also be time-asymmetrical, predicting a privileged direction in time, just like the second law of Thermodynamics does. Finally, in order to preserve free will, it would contain a non-algorithmic element, which means that the future would not be computable from the present. Penrose even believes that Quantum Gravity will explain consciousness.

### *Superconductivity and Entanglement*

For many centuries it was assumed that matter could only assume three states: gas, liquid and solid. It was well known that the same matter can undergo a "phase transition": for example, water can turn into vapor or ice. In 1911 the Dutch physicist Heike Kamerlingh Onnes discovered superconductivity: when certain materials are cooled to temperatures close to absolute zero, they exhibit no electrical resistance, i.e. they transition into the superconducting state. Since then it has become obvious that matter can exist in other forms than the three classical ones.

In 1925 Albert Einstein ("Quantum Theory of a Monoatomic Ideal Gas", 1925) and the Indian physicist Satyendranath Bose discovered that bosons, at similarly extremely low temperature, can form a Bose-Einstein condensate, a superfluid that exhibits the bizarre properties of Quantum Mechanics at a macroscopic level (those properties are usually only experienced at very microscopic levels).

It turns out that the electrons of superconductors do not obey the Pauli principle (according to which there can never be two fermions in the same state at the same time).

In 1957 John Bardeen, Leon Neil Cooper, and John Robert Schrieffer in the USA understood that the electrons of superconductors, instead, bind into pairs, each pair behaving like a boson. All these electron pairs condense in the exact same state of very low energy, just like a Bose-Einstein condensate.

For several decades the only superconductors were achieved at extremely low temperatures, near absolute zero. In 1986 the Swiss physicist Karl Müller and the German physicist Johannes Bednorz discovered the first high-temperature superconductor. This challenged the original explanation of how and why superconductors form.

The Indian physicist Subir Sachdev ("Quantum Criticality", 2011) explained that near a "quantum-critical point" the process of entanglement first described (as impossible) by Einstein creates even more than pairs of electrons: it creates entire populations of entangled electrons that violate the Pauli principle. The electrons can no longer be studied as independent particles.

### *The Trail of Asymmetry*

Somehow asymmetry seems to play a protagonist's role in the history of our universe and our life.

To start with, there would be no universe to talk about if there hadn't been a slight asymmetry between matter and antimatter at the beginning of the universe. And there wouldn't be any interesting structure in the universe if there hadn't been a slight asymmetry in the ripples that created the galaxies.

Cosmological models speculate that the four fundamental forces of nature arose when symmetry broke down after the very high temperatures of the early universe began to cool down. Today, we live in a universe that is the child of that momentous split. Without that "broken symmetry" there would be no electrical force and no nuclear force, and our universe would be vastly impoverished in natural phenomena.

Scientists have also speculated at length about the asymmetry between matter and antimatter: if one is the mirror image of the other and no known physical process shows a preference for either, why is it that in our universe protons and electrons (matter) overwhelmingly prevails over positrons and antiprotons (antimatter)?

Most physical laws can be reversed in time, at least on paper. But most will not. Time presents another asymmetry, the "arrow of time" which points always in the same direction, no matter what is allowed by Mathematics. The universe, history and life all proceed forward and never backwards.

Possibly related to it is the other great asymmetry: entropy. One can't unscramble an egg. A lump of sugar which is dissolved in a cup of coffee cannot become a lump of sugar again. Left to themselves, buildings collapse, they do not improve. Most artifacts require periodic maintenance, otherwise they would decay. Disorder is continuously accumulated. Some processes are irreversible.

It turns out that entropy is a key factor in enabling life (and, of course, in ending it). Living organisms maintain themselves far from equilibrium and entropy plays a role in it.

Moreover, in 1848 the French biologist Louis Pasteur discovered that aminoacids (which make up proteins which make up living organisms) exhibit another singular asymmetry: for every aminoacid there exist in nature its mirror image, but life on Earth uses only one form of the aminoacids (left-handed ones). Pasteur's mystery is still unexplained (Pasteur thought that somehow that "was" the definition of life). Later, biologists would discover that bodies only use right-handed sugars, thereby confirming that homochirality (the property of being single-handed) is an essential property of life.

Finally, an asymmetry presents itself even in the site of thinking itself, in the human brain. The two cerebral hemispheres are rather symmetric in all species except ours. Other mammals do not show preferences for grasping food with one or the other paw. We do. Most of us are right-handed and those who are not are left-handed. Asymmetry seems to be a fundamental feature of our brain. The left hemisphere is primarily used for language and the interplay between the two hemispheres seems to be important for consciousness.

It may turn out to be a mere coincidence, but the most conscious creatures of our planet have also the most asymmetric brains.

Was there also a unified brain at the origin of thinking, whose symmetry broke down later on in the evolutionary path?

### *A Fuzzy World*

Modern physics relies heavily on Quantum Mechanics. Quantum Mechanics relies heavily on the theory of probabilities. At the time, probabilities just happened to fit well in the model.

Quantum Mechanics was built on probabilities because the theory of probabilities is what was available in those times. Quantum Mechanics was built that way not because Nature is that way, but because the mathematical tools available at the time were that way; just like Newton used Euclid's' geometry because that is what Geometry could provide at the time.

Boltzmann's stochastic theories had showed that the behavior of gases (which are large aggregates of molecules) could be predicted by a dynamics which ignored the precise behavior of individuals, and took into account only the average behavior. In retrospect Boltzmann's influence was enormous on Quantum Mechanics. His simplification was tempting: forget about the individual, focus on the population.

Quantum Mechanics therefore prescribed a "population" approach to Nature: take so many electrons, and some will do something and some will do something else. No prescription is possible about a single electron. Quantum phenomena specify not what a single particle does, but what a set of particles do. Out of so many particles that hit a target, a few will pass through, a few will bounce back. And this can be expressed probabilistically.

Today, alternatives to probabilities do exist. In particular, Fuzzy Logic can represent uncertainty in a more natural way (things are not black or white, but both black and white, to some extent). Fuzzy Logic is largely equivalent to Probability Theory, but it differs in that it describes single individuals, not populations.

On paper, Quantum Mechanics could thus be rewritten with Fuzzy Logic (instead of probabilities) without altering any of its conclusions. What would change is the interpretation: instead of a theory about "set of individuals" (or populations) it would become a theory about "fuzzy individuals". In a Fuzzy Logic scenario, a specific particle hitting a potential barrier would both go through and bounce back. To some extent. It is not that out of a population some individuals do this and some individuals do that; a specific individual is both doing this and doing that. The world would still behave in a rather bizarre way, but somehow we would be able to make statements about individuals. However, this approach would allow Physics to return to a science of individual objects, not of populations of objects.

The uncertainty principle could change quite dramatically: instead of stating that we can never observe all the parameters of a particle with absolute certainty, it could state that we can observe all the parameters of a particle with absolute certainty, but certainty not being exact. When I say that mine is a good book, I am being very certain. I am not being exact (what does "good" mean? How good is good? Etc).

The fact that a single particle can be in different, mutually exclusive states at the same time has broad implications on the way our mind categorizes "mutually exclusive" states; not on what Nature actually does. Nature never constrained things to be either small or big. Our mind did.

Any scientific theory we develop is first and foremost a "discourse" on Nature; i.e., a representation in our mind of what Nature is and does.

Some of the limits we see in Nature (i.e., the fact that something is either big or small) are limits of our mind; and conversely some of the perfection that we see in Nature is the perfection of our mind (i.e., the fact that there is a color white or something is cold or a stone is round, while in Nature no object is fully white, cold or round). Fuzzy Logic is probably a better compromise between our mind and Nature, because it allows us to express the fact that things are not just zero or one, white or black, cold or warm, round or square; they are "in between", both white and black, both cold and warm, both...

### *Time: When?*

On closer inspection, the main subject of Thermodynamics, Relativity and Quantum theories may well be Time. Most of the bizarre implications of those theories are things that either happen "in time" or are caused by Time.

Boltzmann interpreted the second law of Thermodynamics (that entropy can never decrease) as basically a definition of time, and, de facto, an unmasking of "time" as an illusion of being alive in a particular time and place. Boltzmann reasoned that the illusion of time is due to the processes of change that we observe. Those processes (especially the most visible ones, such as decay) are driven by the second law of Thermodynamics, which he had proved to be a statistical law of transition from less probable states to more probable states. Thus a living being "feels" the flow of time only because it lives in a world that is transitioning from a less probable state to a more probable state. If life is possible in states of absolute equilibrium, then those living beings would not perceive any flow of time.

Relativity turned Time into one of several dimensions, mildly different from the others but basically very similar to the others. This clearly contrasts with our perception of Time as being utterly distinct from space. Hawking, for example, thinks that originally Time may have just been a fourth spatial dimension, then gradually morphed into a different type of dimension and, at the Big Bang, it became Time as we know it today.

The mathematician Hermann Bondi has argued that the roles of Time are utterly different in a deterministic and in a non-deterministic universe. Whereas in a deterministic universe, Time is a mere coordinate, in a universe characterized by indeterminacy, such as one governed by Quantum Theory, the passage of time transforms probabilities into actualities, possibility into reality. If Time did not flow, nothing would ever be. Things would be trapped in the limbo of wave functions.

The Australian physicist Paul Davies claims exactly the opposite: Time is rather meaningless in the context of a quantum model of the universe, because a general quantum state of the universe has no well-defined time.

With Hawking, Time may not have existed before the Big Bang, and may have originated afterwards by mere accident.

### *Time: What?*
The subject of Time has puzzled and fascinated philosophers since the dawn of consciousness. What is Time made of? What is the matter of Time? Is Time a human invention?

There is no doubt that physical Time does not reflect psychological Time. Time, as we know it, is subjective and relative. There is a feeling to the flow of time that no equation of Physics can reproduce. Somehow, the riddle of Time reminds us of the riddle of consciousness: we know what it is, we can feel it very clearly, but we cannot express it, and we don't know where it comes from.

If you think that there is absolute time, think again. Yes, all clocks display the same time. But what makes you think that what they display is Time? As an example, let's go back to the age when clocks had not been invented yet: time was defined by the motion of the sun. People knew that a day is a day because the sun takes a day to turn around the Earth (that's what they thought). And a day was a day everywhere on the Earth, even among people who had never communicated to each other. Is that absolute Time?

What would happen if the Sun all of a sudden slowed down? People all over the planet would still think that a day is a day. Their unit of measurement would be different. They would be measuring something else, without knowing it. What would happen today if a galactic wave made all clocks slow down? We would still think that ten seconds are ten seconds. But the "new" ten seconds would not be what ten seconds used to be. So clocks do not measure Time, they just measure themselves. We take a motion that is the same all over the planet and use that to define something that we never really found in nature: Time.

At the least, we can say that measurement of Time is not innate: we need a clock to tell "how long it took".

Unfortunately, human civilization is founded on Time. Science, the Arts and technology are based on the concept of Time. What we have is two flavors of Time: psychological time, which is a concrete quantity that the brain creates and associates to each memory; and physical time, an abstract quantity that is used in scientific formulas for the purpose of describing properties of matter.

The latter was largely an invention of Isaac Newton, who built his laws of nature on the assumption of an absolute, universal, linear, continuous Time. Past is past for everybody, and future is future for everybody.

Einstein explained that somebody's past may be somebody else's present or even future, and thereby proved that time is not absolute and not universal. Any partitioning of space-time into space and time is perfectly legal. The only requirement on the time component is that events can be ordered in time. Time is pretty much reduced to a convention to order events, and one way of ordering is as good as any other way.

In the meantime, the second law of Thermodynamics had for the first time established formally the arrow of time that we are very familiar with, the flowing from past to future and not viceversa.

### *Time: Where?*

Once the very essence of Time had been doubted, scientists began to doubt even its existence.

The British physicists Arthur Milne and Paul Dirac are two of the scientists who wondered if the shaky character of modern Physics may be due to the fact that there are two different types of time and that we tend to confuse them. Both maintained that atomic time and astronomical time may be out of sync. In other words, the speeds of planets slowly change all the time in terms of atomic time, although they remain the same in terms of astronomical time. A day on Earth is a day regardless of the speed of the Earth, but it may be lasting less and less according to an atomic clock. In particular, the age of the universe may have been vastly exaggerated because it is measured in astronomical time and astronomical processes were greatly speeded up in the early stages of the universe.

Not to leave anything untried, the US physicist Richard Feynman even argued in favor of matter traveling backwards in time: an electron that turns into a positron (its anti-particle) is simply an electron that turns back in time. His teacher John Wheeler even argued that maybe all electrons are just one electron, bouncing back and forth in time; and so all other particles. There is only one instance of each particle. That would also explain why all electrons are identical: they are all the same particle.

Einstein proved that Time is not absolute and said something about how we experience time in different ways depending on how we are moving. But he hardly explained what Time is. And nobody else ever has.

The British physicist Julian Barbour believes that Time does not exist, and that most of Physics' troubles arise from assuming that it does exist. We have no evidence of the past other than our memory of it. We have no evidence of the future other than our belief in it. Barbour believes that it is all an illusion: there is no motion and no change. Instants and periods do not exist. What exists is only "time capsules", which are static containers of "records". Those records fool us into believing that things change and events happen. There exists a "configuration space" that contains all possible instants, all possible "nows". This is "Platonia." We experience a set of these instants, i.e. a subset of Platonia. Barbour is inspired by Leibniz' theory that the universe is not a container of objects, but a collection of entities that are both space and matter. The universe does not contain things, it "is" things.

Barbour does not answer the best part of the puzzle: who is deciding which "path" we follow in Platonia? Who is ordering the instants of Platonia? Barbour simply points to quantum mechanics, that prescribes we should always be in the "instant" that is most likely. We experience an ordered flow of events because that is what we were designed for: to interpret the sequence of most likely instants as an ordered flow of events.

Barbour also offers a solution to integrating relativity and quantum mechanics: remove time from a quantum description of gravity. Remove time from the equations. In his opinion, time is precisely the reason why it has proved so difficult to integrate relativity and quantum theories.

## *Time: Why?*

In classical and quantum Physics, equations are invariant with respect to time inversion. Future and past are equivalent. Time is only slightly different from space. Time is therefore a mere geometrical parameter. Because of this, Physics offers a static view of the universe. The second law of Thermodynamics made official what was already obvious: that many phenomena are not reversible, that time is not merely a coordinate in space-time.

In the 1970's Prigogine showed, using Boltzmann's theorem and thermodynamic concepts, that irreversibility is the manifestation at macroscopic level of randomness at microscopic level.

Prigogine then attempted a microscopic formulation of the irreversibility of laws of nature. He associates macroscopic entropy with a microscopic entropy operator. Time too becomes an operator, no longer a mere parameter. Once both time and entropy have become operators, Physics has been turned upside down: instead of having a basic theory expressed in terms of wave functions (i.e., of individual trajectories), he obtains a basic theory in terms of distribution functions (i.e., bundles of trajectories). Time itself depends on the distribution and therefore becomes itself a stochastic quantity, just like entropy, an average over individual times. As a consequence, just like entropy cannot be reversed, time cannot: the future cannot be predicted from the past anymore.

Traditionally, physical space is geometrical, biological space (the space in which biological form develops) is functional (for example, physical time is invariant with respect to rotations and translations, biological space is not). Prigogine's Time aims at unifying physical and biological phenomena.

## *Black Holes and Wormholes: Gateways to Other Universes and Time Travel*

By definition, all information about the matter that fell into a black hole is lost forever: a black hole may have been generated by any initial configuration of matter, but there is no record of which one it was.

The Israeli physicist Jacob Bekenstein ("Black Holes and the Second Law", 1972) first conceived that black holes should store a huge amount of entropy. The area of the event horizon is a function of mass, spin and charge. The British physicist Stephen Hawking ("The Singularities of gravitational collapse and cosmology", 1970) had already proven that this area can never decrease, just like the entropy of a closed system can never decrease. The analogy eventually led to an identity: the entropy of a black hole is proportional to the area of its event horizon.

Stephen Hawking ("Black hole explosions", 1974) proved that black holes evaporate, therefore information is not only trapped inside the black hole: it may truly disappear forever. (The time it will take for a black hole to evaporate is proportional to the cube of its mass).

However, later the US physicist Leonard Susskind and the Dutch physicist Gerard 't Hooft ("Dimensional Reduction in Quantum Gravity", 1993) argued that information is probably conserved, i.e. black holes are not information-erasers but information-scramblers. They viewed a black hole as a hologram: information about what has been lost inside the black hole is encoded on the surface of the black hole (in the form of fluctuations of the event horizon).

Bekenstein's and Hawking's studies were particularly relevant because they were the first major attempts to integrate Relativity, Quantum Theory and Thermodynamics.

The disappearance of matter, energy and information in a black hole has puzzled physicists since the beginning, as it obviously violates the strongest principle of conservation that our Physics is built upon. It also highlights the contradictions between Quantum Theory and Relativity Theory: the former guarantees that information is never lost, the latter predicts that it will be lost in a black hole.

Einstein himself realized that black holes implied the existence of a "bridge", originally called "Albert Einstein-Rosen bridge" ("The Particle Problem in the General Theory of Relativity", 1935), between our universe and a mirror universe which is hidden inside the black hole, and in which Time runs backwards. The "wormhole" is a solution to Einstein's own gravitational equations already discovered by the Austrian physicist Ludwig Flamm ("The Foundations of Wave Mechanics", 1916).

The Austrian mathematician Kurt Godel, the same individual who had just single-handedly shattered the edifice of Mathematics, pointed out ("An Example Of A New Type Of Cosmological Solution Of Einstein's Field Equations Of Gravitation", 1949) how Einstein's equations applied to a rotating universe implied that space-time can curve to the point that a particle will return to a previous point in time; in other words, "wormholes" would exist connecting two different points in time of the same universe. John Wheeler, who actually coined the term "wormhole" in 1957, showed that wormholes are born, grow and die (""Causality and Multiply-Connected Space-Time", 1962).

Scientists speculated that two points in space can be connected through several different routes, because of the existence of spatial wormholes. Such wormholes could act like shortcuts, so that travel between the two points can occur even faster than the speed of light.

The New Zealand mathematician Roy Kerr in 1963 and the US physicist Frank Tipler in 1974 found other situations in which wormholes were admissible. In the U.S., Kip Thorne even designed a time machine capable of exploiting such time wormholes. Stephen Hawking came up with the idea of wormholes connecting different universes altogether: Hawking's wave function allows the existence of an infinite set of universes, some more likely than others, and wormholes the size of the Planck length connect all these parallel universes with each other.

### *A brief History of the Universe*

One of the consequences of General Relativity is that it prescribes the evolution of the universe. A few possible futures are possible, depending on how some parameters are chosen. These cosmological models regard the universe as one system with macroscopic quantities. Since the discovery that the universe is expanding in all directions (by the British physicist Edwin Hubble in 1929), the most popular models have been the ones that predict expansion of space-time from an initial singularity, the "Big Bang" (first speculated by the Belgian physicist Georges Lemaitre in 1927). Since a singularity is infinitely small, any cosmological model that wants to start from the very beginning must combine Relativity and Quantum Physics.

The story usually begins with an infinitely small universe, in which quantum fluctuations of the type predicted by Heisenberg's principle are not negligible, especially when the universe was a size smaller than the Planck length.

The fluctuations actually "created" the universe (space, time and matter) after a "Big Bang". Time slowly turned into space-time, giving rise to spatial dimensions. Space-time started expanding, the expansion that we still observe today. In a sense, there was no beginning of the universe: the "birth" of the universe is an illusion. There is no need to create the universe, because its creation is part of the universe itself. There is no real origin. The universe is self-contained, it does not require anything external to start it.

We are not children of a planet or of a star. Tiny random quantum fluctuations in the first fraction of a second after the Big Bang created every structure that we observe in today's universe. We, our planet, our star, our galaxy and everything else are children of tiny random fluctuations.

The universe had a small entropy at the Big Bang. The opposite (high entropy) is a state of thermal equilibrium with matter-energy distributed uniformly at a constant temperature (although with lumps of matter due to gravitation, as Roger Penrose showed,, and in fact the structures of highest entropy are the black holes).

Then the universe expanded. If the mass of the universe is big enough (and this is still being debated, but most cosmologists seem to believe so), then at some point the expansion will peak and it will reverse: the universe will contract all the way back into another singularity (the "Big Crunch"). At that point the same initial argument holds, which is likely to start another universe. For example, John Wheeler speculated that the universe might oscillate back and forth between a Big Bang and a Big Crunch. Each time the universe re-starts with randomly assigned values of the physical constants and laws.

Both the beginning and the end are singularities, which means that the laws of Physics break down. The new universe can have no memory of the old universe, except for a higher entropy (assuming that at least that law is conserved through all these singularities), which implies a longer cycle of expansion and contraction (according to Richard Tolman's calculations).

Some scientists believe that they can remove the singularities. In particular, Hawking has proposed a model in which Time is unbounded but finite, and therefore it is not created in the Big Bang even if the universe today has a finite age. (According to Einstein, space is also finite yet unbounded). In his model, Time emerges gradually from space and there is no first moment.

### *Dark Matter and Dark Energy*

The real conundrum for cosmologists, however, is to find the missing matter: most of the matter required for the universe to be the way it is (to account for the gravity that holds together the galaxies the way they are held together) has never been observed. Physicists are searching for "dark matter" (perhaps 23% of the mass-energy of the universe and five times as abundant as matter) that does not interact with ordinary matter, does not emit and does not reflect light, but whose gravitational effect has been inferred by observing the motion of galaxies. Dark matter cannot be any of the known particles, but it is made of particles.

Furthermore, in 1998 the US physicists Saul Perlmutter, Brian Schmidt, and Adam Riess discovered that the expansion of the universe is not constant, as Hubble thought, but is instead accelerating. In fact, a universe made only of matter would decelerate because of gravitational attraction. It is true that the energy density due to matter decreases as the universe expands, but this decrease is not enough to account for the observed acceleration: there has to be some kind of energy at work that is not due to matter. What is needed is a fixed amount of energy at every point in the universe. This "dark energy" is not associated with particles and it must be about 73% of the mass-energy of the universe in order to account for the observed acceleration. Dark energy is not made of particles. It is not a new particle, otherwise it would be part of matter.

Dark energy must be uniformly distributed throughout space to explain the accelerating universe, whereas "dark matter" must be organized in large chunks surrounding galaxies to explain the existing structures. Dark matter exerts positive pressure (it pulls matter together and creates structures), whereas dark energy exerts negative pressure (it pulls the universe apart).

Unlike matter, which dilutes while the universe expands, "dark energy" is a form of energy that is constant no matter what the universe does. An obvious candidate is vacuum energy, the energy that Heisenberg's principle might create in the vacuum as particles pop up randomly. The anti-gravitational energy that Einstein called "cosmological constant" would be physically equivalent to this vacuum energy. However, this vacuum energy would be many orders of magnitude more than the required amount of dark energy. So then one would have to explain what happens to all the vacuum energy that does not become "dark energy", i.e. that does not contribute to the acceleration of the universe. The Canadian physicist Cliff Burgess ("Extra Dimensions and the Cosmological Constant Problem ", 2007) suggested that vacuum energy may be hidden away in other dimensions, dimensions that are curled and only sensitive to gravitation.

Matter (the sum of "our" matter plus dark matter) may have dominated at the beginning, thus slowing down the expansion. Energy density decreases as the universe expands. After a while, dark energy became predominant and caused the universe to accelerate. Unless some process reverses this trend, eventually matter will be so diluted that its density will be zero and there will only be dark energy to control a universe spinning out of control.

Whatever the reason, more than 90% of the mass-energy of the universe has not been found. Today's Physics is really a scientific theory about the mere 4% of mass-energy that we can account for.

### *The Creation Of Information*

The universe before the Big Bang was in a state of equilibrium. That means minimal information and minimal order. That means maximum entropy. Today it is in a state of non-equilibrium, which implies that entropy declined, which contradicts the second law of Thermodynamics. The solution of the apparent paradox is gravitation: gravitation causes the creation of order (and therefore information).

Gravitation also solves the other puzzle: where did the initial energy come from? It turns out that gravitation behaves like negative energy that equals the positive energy that has created all matter: the two together yield zero energy. There is no need for an initial source of energy to start the Big Bang.

One oddity still remains: the effect of creating the universe does have a visible effect, and that is the many ordered structures that we see in the sky. In other words, the creation of the universe has created an entropy gap. Information has been created that was not there. Where did it come from?

### *The End of Entropy*
Very few people are willing to take the second law of Thermodynamics as a primitive law of the universe. Explicitly or implicitly, we don't seem happy with this law that states an inequality. Somehow it must be a side effect of some other phenomenon.

Thomas Gold (among others) believes that the second law follows the direction of the universe: entropy increases when the universe expands, it decreases when the universe contracts (or, equivalently, when Time flows backwards). The second law would simply be an effect of the expansion or contraction. In that case the universe might be cyclic.

Roger Penrose has also investigated the mystery of entropy. A gravitational effect results in two dual phenomena: a change in shape and a change in volume of space-time. Consequently, Penrose separates the curvature tensor in two components: the Ricci tensor (named after the Italian mathematician Gregorio Ricci who founded the theory of tensors) and the Weyl tensor (named after the German mathematician Hermann Weyl, a close associate of Einstein's). The Weyl tensor measures the change in shape, and, in a sense, the gravitational field, whereas the Ricci tensor measures the change in volume, and, in a sense, the density of matter. The Weyl tensor measures a "tidal" effect and the Ricci tensor measures an effect of volume reduction. The Ricci tensor is zero in empty space, it is infinite in a singularity. The Weyl tensor is zero in the initial singularity of the Big Bang, but infinite at the final singularity of the Big Crunch. Penrose showed that entropy follows the Weyl tensor and the Weyl tensor may hide the puzzling origin of the second law of Thermodynamics.

### *The Resurrection of Information*
The curvature in proximity of a black hole is infinite: all objects are doomed. There is a distance from the black hole which is the last point where an object can still escape the fall: the set of those points defines the horizon of the black hole.

In 1974 Stephen Hawking discovered that black holes may evaporate and eventually vanish. The "Hawking radiation" that remains has lost all information about the black hole. This violates the assumption of determinism in the evolution of the universe, i.e. that, if we know the present, we can always derive the past, because the present universe contains all information about how the past universe was.

Only two options have been found to allow for the conservation of information. The first one is to allow for information to travel faster than light. That would allow it to escape the black hole. But it would violate the law of causality (that nothing can travel faster than light).

The second option is that a vanishing black hole may leave behind a remnant the size of the Planck length. Andrew Strominger has argued for the latter option. This option calls for an infinite number of new particles, as each black hole is different and would decay into a different particle. Strominger believes that such particles are extreme warps of space-time, "cornucopions", that can store huge amount of information even if they appear very small to an outside observer and their information would not be accessible.

After all, Stephen Hawking and Jacob Bekenstein had proved that the entropy of a black hole is proportional to its surface (or, better, the surface area of its event horizon), which means that entropy should decrease constantly during the collapse of the black hole, which means that information must somehow increase, and not disappear...

### *Gravity and Entropy*

Dutch physicist Erik Verlinde ("On the Origin of Gravity and the Laws of Newton", 2010) contends that gravity is just an illusion, or, better, a mere side-effect of the second law of Thermodynamics. That law says that Nature tends to maximize disorder, i.e. that entropy tends to increase. Gravity is an "entropic" force in the sense that it is ultimately the manifestation of differences in entropy.

The US physicist Ted Jacobson had already proved that Einstein's equations of General Relativity can be formulated in terms of statistical properties ("Thermodynamics of Spacetime: The Einstein Equation of State", 1995). The Israeli physicists Ram Brustein and Merav Hadad ("The Einstein equations for generalized theories of gravity", 2010) generalized Jacobson's theory to a wider class of gravitational theories (of which Einstein's General Relativity is just a particular case).

The US physicist Sean Carroll argues that understanding the universe depends on first understanding entropy.

### *Inflation: Before Time*

What was there before the Big Bang created our universe? A widely held "cosmological principle" requires that the universe has no center, no special place. That means that the Big Bang did not occur in a specific point of the universe: it occurred everywhere in the universe, it was the universe. The universe was a point and the Big Bang is merely the moment when it began to expand. By cosmological standards, the Big Bang is still occurring now, in every single point of the universe. Space is being created as the universe expands. There was "nothing" before the Big Bang and there is "nothing" beyond the universe. The Big Bang creates the universe which is everything that exists.

In 1965 the US astronomers Arno Penzias and Robert Wilson accidentally discovered the cosmic microwave background radiation ("A Measurement Of Excess Antenna Temperature At 4080 Mc/s"). This was the remnant of the Big Bang, still travelling across the universe. What puzzled physicist was that it was uniform (constant throughout the universe). This caused two problems. First of all, according to Relativity, nothing can travel faster than light, hence it was impossible for distant regions of the universe to communicate and achieve equilibrium of any sort. Secondly, the universe that we observe is not uniform but has lots of irregularities that we call "galaxies" and the likes.

This "inflationary" model, proposed by Alan Guth ("The Inflationary Universe: A Possible Solution to the Horizon and Flatness Problems", 1980) expanding on the 1948 theory of cosmogenesis by the Ukrainian physicist George Gamow (who, in turn, developed a 1927 idea by the Belgian physicist Georges Lemaitre), assumes that the universe began its life in a vacuum-like state containing some homogeneous classical fields but no particles (no matter as we know it). Then it expanded exponentially (that's the "inflation") and the vacuum-like state decayed into particles. The structure of the universe was created in the first fraction of a second.

Guth's model is based on the existence of scalar fields. A scalar field is one caused by a quantity which is purely numerical, such as temperature or household income. Gravitational and electromagnetic fields, in contrast, also point in a specific direction, and are therefore "vector" fields. Vector fields are perceived because they exert some force on the things they touch, but scalar fields are virtually invisible. Nonetheless, scalar fields play, for example, a fundamental role in unified theories of the weak, strong and electromagnetic interactions. Like all fields, scalar fields carry energy. Guth assumed that in the early stage of the universe a scalar field provided a lot of energy to empty space. This energy produced the expansion, which for a while occurred at a constant rate, thereby causing an exponential growth.

Guth's model solved a few historical problems of cosmology: the "primordial monopole" problem (grand unified theories predict the existence of magnetic monopoles); the "flatness" problem (why the universe is so flat, i.e. why the curvature of space is so small); and the "horizon" problem (how causally disconnected regions of the universe can have started their expansion simultaneously). It does not account for dark matter and energy (for the fact that the exponential inflation slowed down dramatically to what it is now, a thousand times less).

While everybody agrees that the universe is expanding, not everybody agrees on what that means. In the quest for an explanation of dark matter and dark energy, the British physicist Geoffrey Burbidge, the British physicist Fred Hoyle and the Indian physicist Jayant Narlikar developed the "Quasi Steady State Cosmology" (reprised in the "Cyclic Universe Theory" by the US physicist Paul Steinhardt and the British physicist Neil Turok), according to which there is no "Big Bang" to begin with, and there will be no "Big Crunh" to end with. Space and time existed ever since and will exist forever. There is no beginning nor end. The evolution of the universe is due to a series of "bangs" (explosive expansions) and "crunches" (contractions). The Big Bang that we observe today with the most powerful detectors of microwave radiations (first detected in 1964) is simply one of the many expansions following one of the many contractions. Each phase may last a trillion years, and therefore be undetected by human instruments. Burbidge doubts black holes, quasars and the cosmic radiation.

### *Natural Selection for Universes*
Refining Guth's vision, the Russian physicist Andrei Linde came up with a "chaotic inflationary" model ("Nonsingular Regenerating Inflationary Universe", 1982). Linde realized that Guth's inflation must litter the universe with bubbles, each one expanding like an independent universe, with its own Big Bang and its own Big Crunch.

Linde's model is "chaotic" because it assumes a chaotic initial distribution of the scalar field: instead of being uniform, the original scalar field was fluctuating wildly from point to point. Inflation therefore began in different points at different times and at different rates.

Regions of the universe that are isolated by a length greater than the inverse of the Hubble constant cannot be in any relation with the rest of the universe. They expand independently. Any such region is a separate mini-universe. In any such region the scalar field can give rise to new mini-universes.

One mini-universe produces many others. It is no longer necessary to assume that there is a "first" universe.

Each mini-universe is very homogeneous, but on a much larger scale the universe is extremely inhomogeneous. It is not necessary to assume that the universe was initially homogeneous or that all its causally disconnected parts started their expansion simultaneously.

One region of the inflationary universe gives rise to a multitude of new inflationary regions. In different regions, the properties of space-time and elementary particles may be utterly different. Natural laws may be different in each mini-universe.

The evolution of the universe as a whole has no end, and may have no beginning.

The "evolution" of mini-universes resembles that of any animal species. Each mini-universe leads to a number of mini-universes that are mutated versions of it, as their scalar fields are not necessarily the same. Each mini-universe is different, and mini-universes could be classified in a strict hierarchy based on a parent-child relationship.

This mechanism sort of "reproduces" mini-universes in a fashion similar to how life reproduces itself through a selection process. The combinatorial explosion of mini-universes can be viewed as meant to create mini-universes that are ever better at "surviving".

Each mini-universe "inherits" the laws of its parent mini-universe to an extent, just like living beings inherit behavior to an extent through genetic code. A "genome" is passed from parent universe to child universe, and that "genome" contains instructions about which laws should apply. Each genome only prescribes a piece of the set of laws governing the behavior of a universe. Some are random. Some are generated by "adaptation" to the environment of many coexisting universes.

At the same time, expansion means that information is being propagated like in a neural network through the hierarchy of expanding universes.

It may also be that a universe is not born just out of a parent universe, but of many parent universes. A region of the universe expands because of the effect of many other regions. This is similar to what happens with neural networks.

With a little imagination, the view of the chaotic inflationary theory can be interpreted in this way:

- The expansion of a new region may be determined by many regions, not just one.
- Each region somehow inherits its laws from those regions.
- The laws in a region may change all the time, especially at the beginning.
- The laws determine how successful a region is in its expansion.
- Different expansion regions with different laws can communicate. They are likely to compete for survival.
- Adaptation takes a toll on expansion regions. Regions die. Branches of regions become extinct.

Obviously, this scenario bears strong similarities with biological scenarios.

Another theory that presupposes evolving universes is the one advanced by the US philosopher Quentin Smith ("A Natural Explanation of the Existence and Laws of Our Universe ", 1990) and by the US astrophysicist Lee Smolin ("Did the Universe Evolve? ", 1992). He thinks that black holes are the birthplaces of offspring universes. The constants and laws of Physics are randomly changed in the new universes, just like the genome of offspring is randomly mutated. Black holes guarantee reproduction and inheritance. Universes that do not give rise to black holes cannot reproduce: there is therefore also a kind of "natural selection" among Smolin's universes. In this scenario, our universe's delicate balance of constants and forces is the result of evolution.

### *Time Symmetry*

Entropy was low at the beginning (at the Big Bang) and has since been increasing. The destiny of the universe is, apparently, to expand forever until it will be, for all practical purposes, empty, i.e. it will have reached its maximum entropy and a state of permanent equilibrium. Black holes, in particular, have a lot of entropy (the one black hole that dwells in the center of our galaxy has 100 times more entropy than the whole observable universe).

The US physicist Sean Carroll hypothyzed ("Spontaneous Inflation and the Origin of the Arrow of Time", 2004) that the Big Bang may not be the beginning of the universe but merely a state of transition from decreasing entropy to increasing entropy. He noticed that space is not empty because of the very "dark energy" that caused its expansion. This energy causes quantum fluctuations that can create baby universes. The intriguing feature of this scenario is that one can play the universe in both directions of time: there is no arrow of time.

### *Loop Quantum Gravity*

The ultimate goal of Loop Quantum Gravity (LQG) is still the "quantization" of general relativity, but the way it approaches the problem is very different: it is purely geometric.

Roger Penrose ("Applications of negative dimensional tensors ", 1971) toyed with the notion of a "spin network" (derived from Louis Kauffman's "knot theory") in an attempt to explain the structure of three-dimensional space. A spin network is a graph whose edges are labeled by integers, corresponding to the possible values of the angular momentum. Penrose sensed that these could be the simplest geometric structures to describe space. The revolutionary assumption was that space and time are not primitive entities but are themselves constructed out of something more primitive. In quantum terms, space and time can fluctuate, as long as casual sequences are preserved.

The Canadian physicist Bill Unruh ("Notes on black-hole evaporation", 1976) discovered that an accelerating observer must measure a temperature (a black-body radiation) where an inertial observer observes none, the temperature being proportional to the acceleration. This means that the very concept of "vacuum" depends on the state of motion of the observer: an accelerating observer will never observe any vacuum. (This also proved that Einstein's principle of equivalence was slightly incorrect: a constantly-accelerating observer and an observer at rest in a gravitational field are not equivalent, as the former would observe a temperature and the latter would not). Every accelerating observer has a hidden region (all the photons that cannot reach her because she keeps accelerating, getting closer and closer to the speed of light) and a horizon (the boundary of her hidden region).

Jacob Bekenstein's theorem implies that every horizon separating an observer from her hidden region has an entropy. That entropy turns out to be proportional to the information that is hidden or trapped in the hidden region (the missing information). According to Bekenstein's theorem, the entropy of the radiation that the accelerated observer experiences is proportional to the area of her horizon.

Lee Smolin and the Italian physicist Carlo Rovelli ("Knot theory and quantum gravity", 1988) put Unruh and Bekenstein together and realized something that is built into any theory of "quantum gravity" (into any quantization of relativity): the volumes of regions in space must come in discrete units, just like energy comes in discrete units. If energy comes in discrete units, then space must come in discrete units. Just like matter is made of discrete particles, space itself must be made of discrete units. A volume cannot be divided forever: there is an elementary unit of volume.

Smolin used Bekenstein's and Unruh's theorems to prove that spacetime must be discrete. If spacetime were continuous, then a volume of spacetime (no matter how small) would contain an infinite amount of information. But for any volume of spacetime an accelerating observer would observe a finite entropy (finite because it is proportional to the surface of the volume) and therefore a finite amount of missing information. The amount of information is finite because the surface of the horizon is finite and therefore entropy is finite. The amount of information within a volume of spacetime must be finite, therefore spacetime cannot be continuous. Spacetime must have an "atomic" structure just like matter has an atomic structure. (This conclusion had been reached independently by Jacob Bekenstein in his studies on the thermodynamics of black holes).

Kenneth Wilson had first hypothesized that space was a discrete lattice ("Confinement of Quarks", 1974). What Smolin did was to make Wilson's discrete lattice also change dynamically, able to evolve in time, as General Relativity requires. In his formulations the inter-relationships among Wilson's structures (the "loops") define space itself. Smolin used the work of two Indian scientists. Abhay Ashtekar ("New Variables for Classical and Quantum Gravity ", 1986) came up with the "loop-space model", based on a theory by Amitaba Sen ("Gravity as a Spin System", 1982) that splits time and space into two distinct entities subject to quantum uncertainty (analogous to momentum and position). The solutions of Einstein's equations would then be quantum states that resemble loops. Smolin's theory was simply a theory of loops and how they interact and combine. (The Uruguayan physicist Rodolfo Gambini had independently reached similar conclusions).

In this way Einstein's theory of gravitation (General Relativity) is reformulated to resemble Maxwell's theory of electromagnetism, with loops playing the role of field lines.

Loop-quantum gravity has dramatic implications on cosmology. For example, the Big Bang turns out to be a "Big Bounce" from an imploding universe to an expanding universe. The world as we know it with all its galaxies, stars and moons was created by tiny ripples in spacetime (otherwise the universe would be a boring homogeneous lattice). The Big Bounce contains quantum fluctuations that would explain those tiny ripples and therefore our galaxy, sun, planet and, ultimately, myself typing these words.

Loop-states turned out to be best represented by Penrose's spin networks. The lines of a spin network carry units of area. The structure of spin networks generates space.

The space that we experience is continuous. Spin networks, instead, are discrete. They are graphs with edges labeled by spins (that come in multiples of 0.5) and with three edges meeting at each vertex. As these spin networks become larger and more complex, they "yield" our ordinary, continuous, smooth 3-dimensional space. A spin network, therefore, "creates" geometry. It is not that a spin network yields a metric (the metrics being what define the geometry of a region of space) but that each vertex of a spin network creates the volume of a region of space.

An evolving spin network (a "spin foam") is basically a discrete version of Einstein's spacetime. Spin-foams are four-dimensional graphs that describe the quantum states of spacetime, just like spin networks describe the quantum states of space. Spin foams describe the quantum geometry of spacetime (not just space). A spin foam may be viewed as a quantum history. Spacetime emerges as a quantum superposition of spin foams (topologically speaking, it is a two-dimensional "complex").

The way spin networks combine to form space is not clear, as there seems to be no "natural law" (no equivalent of gravitation or of electromagnetism) at work. Spin networks "spontaneously" combine to form space. The formation of space resembles the Darwinian process that creates order via natural selection of self-organizing systems. Space appears to be the result of spontaneous processes of self-organization à la Stuart Kauffman.

The hypothesis that space is discrete also helps remove some undesired "infinites" from Quantum Theory. For example, charged particles interact with one another via electromagnetic fields. The electromagnetic field gets stronger as one gets closer to the particle. But a particle has no size, so one can get infinitely closer to it, which means that the field will get infinitely strong. If space is discrete instead of continuous, the paradox is solved: there is a finite limit to how close to a particle one can get.

Spin networks thus solve "quantum gravity" in three dimensions. Spin networks describe the quantum geometry of space. In order to introduce the (fourth) temporal dimension, a concept of "history" has been added by some researchers.

Basically, Einstein's great intuition is that spacetime is not just a stage but an actor in the story of the universe, a story that evolves from its interaction with matter, and Loop Quantum Gravity shows that it is made of its own atoms just like matter is made of its own atoms.

The US physicist Martin Bojowald ("Loop Quantum Cosmology", 2008) deduced that a true singularity cannot exist, as each space atom can only contain a finite amount of energy-matter. Therefore he revised Big Bang cosmology: the Big Bang was a moment of maximum density that must have come from a previous Big Crunch. In other words, the story of the universe should be roughly symmetric (or, better, a mirror image) before and after the Big Bang: there was a collapsing universe before the Big Bang created an expanding universe.

The Greek physicist Fotini Markopoulou showed that spin networks evolve in time in discrete steps: at every step, the change of each vertex of the spin network only depends on its immediate neighbors. This is reminiscent of Von Neumann's cellular automata and of algorithm-based thinking, as opposed to the traditional formula-oriented thinking of Physics.

Markopoulou ("The Internal Description of a Causal Set", 1999) introduced causality in Loop Quantum Gravity. In her view, time is not an illusion, just an approximation. She compares it to the river that seems to flow in a smooth way even though the motion of its water molecules is chaotic. Causality does exist at a very fundamental level, although it may not be the one that we perceive in our daily life.

The idea of Loop Quantum Gravity was further expanded by Causal Dynamical Triangulation, a theory introduced in the 1990s by Renate Loll (Dutch), Jan Ambjorn (Danish) and Jerzy Jurkiewicz (Polish) that constructs spacetime from elementary structures called "four-simplexes". A four-simplex is the four-dimensional equivalent of a tetrahedron.

Yet another approach to quantum gravity is purely mathematical. For example, Markopoulou noticed similarities between the "categories" used by General Relativity and those used by Quantum Theory. These have little in common with the traditional category of Physics whose objects are sets and whose morphisms are functions. In her view Quantum Theory is better understood as a theory of spacetime.

### *Euclidean Quantum Gravity*

"Euclidean Quantum Gravity" is a term that refers to the idea that space and time are treated as equals, and that spacetime at any point in time is the superposition of all the possible shapes of spacetime. This model is wildly unstable. However, the Danish physicist Jan Ambjorn and the German physicist Renate Loll ("Non-perturbative Lorentzian Quantum Gravity, Causality and Topology Change", 1998) removed the "Euclidean" clause and introduced the time arrow in the building blocks (or "simplices") of spacetime. In that case the building blocks tend to assemble themselves in the kind of spacetime that we observe (notably, the four dimensions). Basically, a four-dimensional spacetime emerges spontaneously through a process of self-organization similar to the one that yields crystals and many biological systems. The catch is that all building blocks (simplices) must share the same arrow of time (i.e., causality must be encoded at the smallest level of organization).

The Czech physicist Petr Horava ("Quantum gravity at a Lifshitz point", 2009) proposed to solve the contradictions of Quantum Physics and General Relativity by separating space from time at high energy. Basically, at high energy Einstein's spacetime would decompose into Newton's separate dimensions for space and time. The idea is intuitive enough: General Relativity is about gravitation, which is a low-energy phenomenon, whereas Quantum Physics is about particles and waves, which are high-energy phenomena. Each one has a preferred "domain" of competence. Horava's theory simply finds a compromise between the two, assuming that one emerges from the other depending on the level of energy.

## *The Holographic Principle*

Inspired by the fact that the entropy of a black hole is proportional to its surface and believing that the Planck length is one side of an area that can hold only one bit of information, Gerard t Hooft ("Cosmology in 2+1 dimensions", 1993) generalized those ideas and proposed that the informational content of a region of space can always be equivalently expressed by a theory that lives on the boundary of that region (the so called "holographic principle"). Everything that happens in a three-dimensional space is encoded in a two-dimensional surface.

In particular, the horizon of a black hole (a two dimensional surface) stores all the information that ever fell into the hole (the solution to Hawking's paradox of information evaporation).

The Argentine physicist Juan Maldacena ("The Large N Limit of Superconformal Field Theories and Supergravity", 1997) realized that one could represent a universe described by superstring theory functioning in an anti-DeSitter spacetime (a negatively curved spacetime) with a quantum field theory operating on the boundary of that spacetime. And viceversa.

In other words, one could imagine a two-dimensional universe with no gravity that exists on the boundary of a three-dimensional universe with gravity. The two universes are equivalent. The three-dimensional universe which we perceive might indeed be encoded on a two-dimensional surface, like a hologram. This makes spacetime less "fundamental" than assumed if it can be reduced to something simpler.

## *Gravity As The Vacuum*

The striking difference between the gravitational force and the other three forces is that the gravitational force is much weaker. Gravitational force operates among "bodies" (e.g., complex aggregates of interacting particles), but does not seem to have any effect on interactions between elementary particles.

The effects of the gravitational force are not considered in the Standard Model because they are assumed to be negligible among subatomic particles, but, of course, the other rational explanation would be that... there are no such effects on subatomic particles: based on empirical evidence, one could assume that gravitation does not operate at all among elementary particles but only among bodies made of many interacting particles.

Physicists had known for a long time that there is a superficial similarity between Einstein's equation of how the momentum-energy tensor affects the curvature of space-time and Maxwell's equation of the electromagnetic field, but the connection between gravitation and electromagnetism was made only in 1968, and it had to do with the vacuum.

The Russian physicist Andrei Sakharov ("Vacuum Quantum Fluctuations in Curved Space and the Theory of Gravitation", 1968) proposed a simple explanation to these "oddities": that gravitation might be not a fundamental interaction but a by-product of the electromagnetic interaction, precisely an electromagnetic phenomenon induced by the presence of matter in the quantum vacuum (the quantum field that is present even in "empty" space). Thus, according to Sakharov, matter is not just "there" but is "in" the quantum vacuum, and therefore interacts with it, causing some kind of quantum-fluctuation energy. That fluctuation is gravitation. In a sense, Sakharov moved to Quantum Theory the same objection that Mach had moved to Newton's Physics: you can't assume that a body is isolated, because a body is always "interacting" with something. Except that Sakharov updated this idea to the terminology of Quantum Theory. A body, any body, is immersed in quantum fields, and thus interacts with them. One such field is the quantum vacuum. Sakharov thought that this was not negligible at all, in fact that it was the origin of the gravitational interaction itself.

Building on Sakharov's ideas, the US physicist Harold Puthoff ("Gravity as a Zero-Point-Fluctuation Force," 1989) proposed that a body's inertia (as expressed by Newton's equation "F=ma") is due, as Mach speculated, to the distribution of matter in the universe, and, more precisely, to the electromagnetic interaction that arises from quantum fluctuations of the zero-point field in accelerated frames. Basically, a particle's inertia is a function of the particle's interaction with the zero-point field. Inertia is resistance to acceleration: Puthoff showed that the resistance which defines the inertia of a particle is, ultimately, electromagnetic resistance caused by the zero-point field on the particle. Matter (made of leptons and quarks) continuously interacts with the zero-point field, and this interaction yields a force (the "resistance" to motion) whenever acceleration takes place. Inertia is due to the distortion of the zero-point field under acceleration. Newton's equation "F=ma" is merely an effect of zero-point fluctuations in an accelerated reference frame.

Puthoff has, de facto, offered the first rational explanation of why force and acceleration are related; and has also given a definition of what "mass" is (Newton introduced it as a primitive property of matter). Technically, inertia is due to the high frequencies of the distortion of the zero-point spectrum, whereas gravity is due to its low frequencies. Newton and Mach measured inertia against a reference frame. Puthoff is using the zero-point field as the reference frame, and is claiming that inertia originates from the reference frame itself.

The "gravitational field" is the set of all electromagnetic fields generated by all particles (leptons and quarks) as they interact with the zero-point field.

The Standard Model has been able to predict correctly all sorts of complicated experiments. However, a skeptic would object that the Standard Model does not quite explain everything that we see in nature (falling rocks, flying birds or suspended bridges). It only explains phenomena that quantum physicists perform in laboratories and that cosmologists presume are happening billions of light-years away. Newton's Physics does explain the phenomena of the human world (the world of objects). It had been relatively easy to reduce Relativity Theory to Newton's Physics, but it had been very difficult to reduce Quantum Theory to Newton's Physics.

## *Spacetime as Entanglement*

Juan Maldacena spawned a new view of spacetime when he showed that quantum entanglement is key to the existence of space-time ("The Large N Limit of Superconformal Field Theories and Supergravity", 1998).

The Japanese physicists Shinsei Ryu and Tadashi Takayanagi showed that quantum entanglement and spacetime seem to be dual concepts, and that spacetime may simply be an emergent property of quantum entanglement ("Holographic Derivation of Entanglement Entropy", 2006).

Mathematically, the US physicist Brian Swingle proved that tensor networks connect entanglement to the geometry of spacetime ("Entanglement Renormalization and Holography", 2012). Swingle showed that higher-degree entanglement behaves just like distance in space: the degree of entanglement is de facto a distance through an extra spatial dimension. Swingle and the Canadian physicist Mark Van Raamsdonk showed that the universality of gravity is directly related to the universality of entanglement ("Universality of Gravity from Entanglement", 2014).

Susskind and Maldacena showed that two ideas proposed by Einstein and Nathan Rosen in 1935, namely quantum entanglement and wormholes, might be closely related ("Cool Horizons for Entangled Black Holes", 2013): the radiation emitted by a black hole (the one predicted by Hawking) remains forever entangled with the inside of the black hole via multiple microscopic wormholes.

## *The Origin of Mass*

Newton's mass was an arbitrary concept. Einstein tried to explain it in terms of energy. Quantum Chromodynamics suggests that most of what we call "matter" is not all that material, because a body is made of elementary "particles" that are almost mass-less (for example, the proton is made of two quarks, whose combined masses are about 1% of the mass of the proton, and of gluons, which are mass-less). Newton believed that mass could not be created: it was always conserved. Einstein showed that energy can turn into mass and viceversa. That opened the possibility that mass might not be a primitive element, but might be "created"; that there is an origin of mass. The US physicist Frank Wilczek visualizes the origin of mass in a compromise that Nature has to strike between two opposing principles: on one hand Nature wants a quark and an anti-quark to be as near as possible to minimize the energy required (the strong/color force increases with distance) but pinpointing an anti-quark's position next to its quark would require an infinite amount of energy (as per Heisenberg's uncertainty principle); and viceversa (the energy is minimal when the two particles are let loose in the universe, but then the strong force between them would become infinite). The compromise between these two extremes is the mass of the proton.

Noting that photons become heavy inside (electric) superconductors (or, better, that an observer inside a superconductor would perceive a photon as a massive particle), Wilczek derives the analogy that we live inside a (non-electric) superconductor within which particles (and then objects) acquire mass. That "superconductor" is made of the Higgs condensate, which is made from the Higgs particle.

This model seems to point towards a deeper truth. Since Quantum Physics predicts that empty space is inherently unstable, both Newton's absolute space and Einstein's relative spacetime (that defied common sense by assuming action at a distance) are being replaced by the vision of a space that is full of spontaneous activity, that seems to have a life of its own. In fact, there seems to be layers of activity that take place in spite of the laws of Physics at our layer. That activity is not irrelevant: it might determine what happens at our layer. These "condensates" arise from empty space, which behaves like a superconductor of a new kind. One theory is that the metric field of General Relativity should also obey the same laws, and therefore be "inhabited" by quantum fluctuations, and be made of yet another kind of "condensate". After all, Einstein's cosmological constant (if it is indeed needed) would de facto represent an intrinsic property of space.

### *Structural Realism*

The British philosopher John Worrall introduced an interpretation of Quantum Theory, inspired by Henri Poincaré, that takes "relations" as the basic units of reality ("Structural Realism", 1989).

Particles are not good building blocks because, according to Quantum Theory, they are not localized: in general, there is no specific place where a particle is. In fact, even if one managed to localize a particle in a specific point of space, a moving observer might see that particle spreading over the entire universe. Particles move all the time, but they don't have clear trajectories. The very number of particles within a system depends on who counts them. Laboratory instruments don't really detect particles as much as interactions between the instrument and some invisible event.

Fields are no more useful building blocks for reality because they too are misnomers in Quantum Theory: a quantum field is not like a classical field but more like an abstract mathematical concept that needs to be coupled with another abstract mathematical concept (a "state vector" or "wavefunction" describing the configuration of the system) in order to make any prediction about a system that happens to be inside the field.

Structural realism confines itself to the study of relations among the constituents of nature ("epistemic structural realism") or even takes relations as being the very constituents of nature ("ontic structural realism"). Combining relations yields structures. The human eyes translates those structures into visible objects, but all there really is (or all that we can really measure) is the relations that create those structures. We cannot know what entities engage in those relations, or those entities simply don't exist and only relations exists as primitive entities.

The German philosopher Meinard Kuhlmann introduced a variant of structural realism in which the basic elements are properties. All we know and all we can measure are properties of what we call "objects". An "object" is simply a bundle of properties. It turns out that this is actually the way we perceive the world in the first year of our life.

Of course, one could also claim the "instrumentalist" view that all scientific theories are simply mathematical abstractions to make predictions, and they do not necessarily explain the real nature of nature.

### *Symmetry and Chaos*

Humans have always assumed that Nature must be fundamentally elegant. This was the guiding principle for Galileo, Newton, Maxwell and Einstein, who in fact came up with elegant equations. "Elegant" in Physics does not mean what it means in art and fashion: it ultimately means "symmetric". The idea is that the universe must have been created symmetric from the beginning, and Physics is really just about discovering the original symmetry from which our highly un-symmetric universe arose. Einstein's General Relativity was, ultimately, about general covariance, the most general kind of symmetry: the equations must have the same form in any system of coordinates.

Therefore it is not surprising that scientific theories were reformulated in terms of symmetries. A symmetry was easy to find in the electromagnetic force because there are both positive and negative electrical charges. A similar symmetry was found in the strong force. Both are of the "mirror" kind. Within these forces an experiment cannot be distinguished from its mirror image. The weak force, however, refuses to comply: a mirror image behaves differently. The three forces were eventually unified by abandoning the requirement of mirror symmetry: it turns out that the mirror symmetry of those two forces was a sheer accident of nature. Nonetheless the unified theory of those three forces prescribed a broader kind of symmetry. This higher-level symmetry is "spontaneously broken" to yield the more limited electromagnetic symmetry which does not apply to the weak force.

This became a general idea. There must be a supersymmetry that encompasses everything in the universe, and all the limited symmetries that we encounter must be due to some kind of "spontaneous symmetry breakdown".

Unfortunately, the higher-level symmetry predicts the existence of a number of mass-less particles that turn out to have mass. Hence, spontaneous symmetry breaking is not enough to explain the oddities of our universe. A mechanism must also assign mass to these particles, and that's Peter Higg's hypothesis of a particle (the Higgs boson) that bestows mass on other particles. The Higgs field is supposed to pervade every corner of the universe. Without the Higgs field, all particles would have no mass and travel at the speed of light. One could imagine that, instead of a Big Bang, the universe as we know it was created when a swarm of particles, traveling at the speed of light, entered the Higgs field and was slowed down by it, thereby acquiring masses and becoming the galaxies, planets, rivers, trees and insects that we are familiar with.

Andrei Linde's chaotic inflation provides a credible model. Linde's starting point is a completely chaotic universe: there is no symmetry at all. It was only by accident that some regions of the universe ended up being uniform enough that a positive feedback within them caused them to expand rapidly and create uniform universes. The inhabitants of each of these universes observe regular (symmetric) laws of nature, their symmetry due to the accidental orientation that started its expansion. The more each universe expands, the "smoother" it appears to be, just like a deflated balloon appears to be a contorted shape but it becomes a perfect sphere when inflated. We live in one such universe that appears to us to have uniform, symmetric laws of nature, while in reality that symmetry is a mere accident, a mere illusion. No wonder therefore that it is "broken".

Another path to the similar conclusions was presented by the British physicist Frank Close. He viewed the quantum vacuum (filled with zero-point energy and countless "virtual" particles) as a medium. Just like any other medium (say, water) it undergoes phase changes. The life of the vacuum then becomes the essential driving force for everything that we observe in our macroscopic lives. The history of the universe than becomes the history of the vacuum. One property to emerge from the vacuum during such a phase transition is the Higgs field, which gave and still gives mass to particles. The universe is one giant quantum fluctuation of the vacuum. And, of course, this is just one such fluctuation. There might be many others that yielded many other universes.

The one law that all these universes have in common is Einstein's field equation: it is that equation that drives their expansion. That equation describes the ultimate layer of reality.

### *The Mathematical Universe*

Eugene Wigner had famously argued ("Symmetries and Reflections", 1967) that "the enormous usefulness of mathematics in the natural sciences is something bordering on the mysterious". In particular, symmetry laws provide human minds with a set of "super-laws" (sort of Kant-ian transcendental principles) that allow us to discover the laws of nature. These symmetries (which, per se, simply express the inability to distinguish physical situations) set the world in motion because they eventually express dynamic relationships as well.

The US mathematician Max Tegmark ("Is the Theory of Everything merely the ultimate ensemble theory?", 1996) argues against the Copenhagen interpretation of Quantum Theory (that there is no reality without observation). He prefers to ground his version of Quantum Theory into the assumption that there exists an external reality independent of human minds. The implication is that this reality must be perceivable also by non-human minds. He can only come up with one kind of external reality that would be perceived identically by all kinds of minds: a mathematical structure. Therefore he argues that the ultimate reality of the universe (its external physical structure) is a mathematical structure. We are, in a sense, thinking equations within a complex system of equations. This would also be the simplest explanation of Wigner's paradox.

No surprise then that the standard model of Quantum Theory is represented by a symmetry: $SU(3) \times SU(2) \times U(1)$

In Tegmark's interpretation the universe neither started nor was created: it simply is (it is a mathematical structure that has dynamic implications).

Meanwhile, a school of "quantum information theory" was revisiting the universe as made of information. The US physicist Seth Lloyd conceived the properties describing a particle as information ("Black Holes, Demons and the Loss of Coherence", 1988). When two particles become entangled, the information that describes each of them individually "declines", whereas the information that describes the pair as a system increases. Particles became increasingly entangled with one another. In parallel, quantum uncertainty evolves through quantum entanglement. A particle gradually loses its quantum identity and becomes part of a quantum collective state. This process continues until a state of equilibrium is reached, that state being one in which the individual particle contains no information and the system contains all the information. Basically, he discovered a universal tendency of systems towards equilibrium thanks to quantum entanglement that inevitably de-personalizes its particles and creates a stronger identity for the whole. In other words, quantum entanglement causes a loss of information, and this loss of information drives a system into equilibrium.

Others, like the German physicist Peter Reimann ("Foundation of Statistical Mechanics under experimentally realistic conditions", 2008) and the British physicist Anthony Short ("Quantum Mechanical Evolution Towards Thermal Equilibrium", 2009), rediscovered the idea and realized that "time's arrow" can be explained by focusing on how information shifts from individuals to ensembles as the quantum waves of the individuals become more and more entangled.

### *Ripples in Spacetime*

In 2005 the Italian mathematician Piero Scaruffi proposed an approach to reconciling Quantum Theory and Relativity Theory that is consistent with Linde's chaotic inflationary theory. Relativity prescribes a spacetime continuum, whereas Quantum Theory prescribes a discrete world. If one views Quantum Theory as a special case of Relativity Theory, this might make sense.

Quantum Theory cannot explain why nature only likes some discrete values.

Relativity Theory is apparently stating some grander and somewhat more absolute truth about the universe; basically, about the dimensions of existence. Quantum Theory is apparently telling us something about the human world of objects and measurements; basically, about the world of "sizes".

One way to visualize these ideas is to think of relativistic spacetime as an ocean, and of quantum values as the ripples caused by an object moving through spacetime. The ocean is a continuum, but the ripples are discrete. Both the ocean and the ripples are real, and one can construct a theory to describe the ocean and a different theory to describe the ripples. Then Relativity is the theory about the ocean, and Quantum Theory is the theory about the ripples.

Quantum Theory, basically, describes the ripples in spacetime (or at least in the region of spacetime inhabited by human observers) caused by energy-matter in motion. Einstein's equations describe how spacetime warps because of matter. Schroedinger's equations describe the ripples caused by such matter.

If this is correct, Relativity's space and time are different from Quantum Theory's space and time. We use the same name for two different things: Relativity's space is a dimension, an underlying framework, whereas Quantum Theory's space is about the "size" of an object. Spacetime is the continuum that energy-matter interacts with. Quantum values are the results of measuring the ripples caused by that interaction. The reason that the ripples are discrete and not a continuum is exactly the same as for ripples that form on a surface. Any attribute of an object over the ripples admits only some values, because it exists and is measured only over the ripples.

Scaruffi believes that the probabilistic nature of Quantum Theory emerges because of the translation from "ocean" to "ripples". Ditto for the attributes (charge, spin, etc). They are all manifestations of the ripples.

General relativity is about the dynamics of the universe: it describes an eternal dance between the distribution of masses and the geometry of spacetime, the former determining the latter and the latter determining the motion of the former. Indirectly, Einstein's equations can be interpreted as the relationship (or the connection" of everything to everything else. As connections change, other connections change. Each connection gets adjusted so as to preserve some global feature of the universe. Basically, the universe behaves like a giant brain.

The quantization of spacetime is a superficial consequence of the "connection of everything to everything else", as this is equivalent to constructing a landscape of attractors.

### *Brains, Lives, Universes*

Let's take a closer look at Life. We have organisms. Each organism is defined by its genome. An organism's genome does not vary during its lifetime. The genome of its offspring varies. The variation is the result of random processes. Each organism interacts with the environment and may or may not survive such interactions. Indirectly, interactions with the environment determine how genomes evolve over many generations.

Then we have neural networks. The behavior of each thinking organism is controlled by a neural network. The principle of a neural network is that of interacting with the environment, propagating the information received from the environment through its neurons and thereby generating behavior. Each neuron has influence over many neurons and what determines the behavior are the connections between neurons. A neural network changes continuously during the life of an organism, especially at the very beginning.

Within neural networks a selection process also applies. Connections survive or die depending on how useful they are. Connections are stronger or weaker depending on how useful they are. Usefulness is defined by interaction with the environment.

Genomes and neural networks are systems that have in common the principle of propagating information about the environment within itself through a process of a) interaction with the environment, b) feedback from the environment, c) selection.

Neural networks, genetic algorithms and chaotic inflationary universes seem to obey very similar principles. They "expand" in order to

- Propagate information within the individual, so as to determine behavior
- Propagate information within the population, so as to determine evolution

### *The Nature of the Laws of Nature*

Even with the sophistication of Relativity Theory, our universe presents us with an uncomfortable degree of arbitrariness.

What is still not clear is why laws (e.g. Einstein's field equations) and constants (e.g., the Planck distance) are the way they are. Why is the universe the way it is?

In 1916 the German physicist Arnold Sommerfeld introduced a quantity that he called "fine-structure constant" that combined all the main constants of Physics: the speed of light, the electric charge of the electron, Planck's constant and the vacuum permittivity. The universe behaves the way it behaves because the value of the fine-structure constant is 1/137. Why 137? (Physicists such as John Barrow and John Webb actually believe that this constant changed over the course of our universe's life, and our universe is but one of many universes, each with a different value for the fine-structure constant).

Furthermore: why do properties of matter such as electrical charge and mass exert forces on other matter? Why do things interact at all?

The most popular cosmological models presume that the physical laws we know today were already in effect at the very beginning, i.e. were born with the universe, and actually pre-existed it. The laws of Physics are simply regularities that we observe in nature. They allow us to explain what happened, and why it happened the way it happened. They also allow us to make predictions. Science is all about predictions. If we couldn't make predictions, any study of Nature would be pretty much useless. We can build bridges and radios because we can make predictions on how things will work.

Three aspects of the fundamental laws are especially puzzling.

The first has to do with the nature of the laws of Nature. How absolute are they? Some laws can be reduced to other laws. Newton's law of gravitation is but a special case of Einstein's. It was not properly a law of Nature, it was an effect of a law of nature that Newton did not know. These days, we are witnessing a quest for a unification theory, a theory that will explain all four known forces (weak, nuclear, electric and gravitational) in one "megaforce": if the program succeeds, we will have proved that those four forces were effects, not causes. Is the second law of Thermodynamics a law indeed, or just the effect of something else?

After all, the laws as we study them today in textbooks are the product of a historical process of scientific discovery. Had history been different (had progress followed a different route) we may have come up with a description of the universe based on different laws, that would equally well fit (individually) all the phenomena we are aware of.

The second question is why are they mathematical formulas. Mathematics is a human invention, but it is amazing how well it describes the universe. True, Mathematics is more a discovery process than an invention process. But, even so, it is a discovery of facts that occur in the realm of mathematical ideas (theorems and the likes). It is amazing that facts occurring in that abstract realm reflect so well facts that occur in the physical realm.

Most Mathematics that is employed today so effectively for describing physical phenomena was worked out decades and even centuries before by mathematicians interested only in abstract mathematical problems. The rule almost never fails: sooner or later a physical phenomenon will be discovered that perfectly matches a mathematical theory. It feels like the universe is a foreign movie, subtitled in mathematical language.

Even more intriguing is the fact that the world of Mathematics is accessible by the human mind. Our bodies have privileged access to physical space, our minds have privileged access to the notes that describe it. We get both treats. The body perceives physical reality through the senses, the mind perceives mathematical reality through reasoning.

The third question is whether they are truly eternal. Were they always the same? Will they always be the same?

Naturally, if the answer is negative, then we don't know anything.

It would seem more likely that they are part of the universe and therefore came to be precisely when the universe came to be. In that case it would therefore be impossible to compute a model of how the universe was born, because we don't know which laws (if any) were in place before the universe was born!

(We don't even know for sure whether the laws of Nature are the same in the whole universe. We don't even know if they have been the same all the time or if they have been changing over time).

Similar arguments hold for the "constants" of Physics, for the dimensionless parameters that shape the laws of nature, in particular for the speed of light, the Planck constant, and the charge of the electron. Why do they have the value they have? Einstein asked: did God have a choice when he created the universe? Could those numbers be different, or are they the only combination that yields a stable universe? A famous formula has been puzzling scientists: the square of the charge of the electron divided by the speed of light and by the Planck constant is almost exactly 1/137. Why?

We don't have a science of natural laws which studies where laws come from. Laws are assumed to transcend the universe, to exist besides and despite the existence of a universe. But that's a rather arbitrary conclusion (or, better, premise).

### *The Nature of Spacetime*

According to Relativity, everything sits in spacetime. According to Relativity, spacetime is not a substance. Sound waves require a medium to travel in (and their speed depends on the medium), but light waves do not require any medium (and it is always the same). According to Einstein, light travels in vacuum. There is no need for an "ether".

The Canadian physicist William Unruh proved the stunning similarity between sound waves propagating in moving fluid and light waves propagating in curved spacetime ("Experimental Black-Hole Evaporation?", 1981). Basically, a black hole is just one example of a class of physical systems that generate an event horizon from which it is impossible to "return". The US physicist Ted Jacobson speculated that spacetime might indeed be a "substance", a kind of fluid, and be discrete at the microscopic scale ("Trans-Planckian redshifts and the substance of the space-time river", 1999).

### *The Prodigy of Stability*

Chaos is a matter of life in this universe. What is surprising is that we do not live in chaos. We live in almost absolute stability. The computer I am writing on right now is made of a few billion particles of all kinds that interact according to mechanic, gravitational, electric, magnetic, weak and strong forces.

The equations to describe just a tiny portion of this computer would take up all my life. Nonetheless, every morning I know exactly how to turn on my computer and every day I know exactly how to operate on it. And the "stability" of my computer will last for a long time, until it completely breaks down. My body exhibits the same kind of stability (for a few decades, at least), so much so that friends recognize me when they see me and every year the IRS can claim my tax returns (no quantum uncertainty there).

Stability is what we are built to care for. We care very little about the inner processes that lead to the formation of a tomato plant: we care for the tomatoes. We care very little for the microscopic processes that led a face to be what it is: we care for what "it looks like". At these levels stability is enormous. Shape, size, position are stable for a number of days, weeks, months, maybe years. Variations are minimal and slow. The more we get into the detail of what we were not built to deal with, the more confused (complex and chaotic) matter looks to us, with zillions and zillions of minuscule particles in permanent motion.

Science was originally built to explain the world at the "natural" level. Somehow scientists started digging into the structure of matter and reached for lower and lower levels. The laws of Physics got more and more complicated, less and less useful for the everyman.

Even more surprising, each level of granularity (and therefore complexity) seems largely independent of the lower and higher levels. Sociology doesn't really need Anatomy and Anatomy doesn't really need Chemistry and Chemistry doesn't really need Quantum Theory.

The smaller we get the more the universe becomes messy, incomprehensible, continuously changing, very unstable. We have grown used to thinking that this is the real universe, because the ultimate reduction is the ultimate truth.

The surprising thing is that at higher levels we only see stability. How does chaos turn into stability? We witness systems that can create stability, order, symmetry out of immense chaos.

One answer is that maybe it is only a matter of perception. Our body was built to perceive things at this level, and at that level things appear to be stable just because our senses have been built to perceive them stable. If our senses weren't able to make order out of chaos, we wouldn't be able to operate in our environment.

Another answer, of course, could be that all other levels are inherently false...

### *A Self-organizing Universe*

The main property of neural networks is feedback: they learn by doing things. Memory and learning seem to go hand in hand. Neural networks are "self-organizing" objects: response to a stimulus affects, among other things, the internal state of the object. To understand the behavior of a neural network one does not need to analyze the constituents of a neural network; one only needs to analyze the "organization" of a neural network.

Physics assumes that matter has no memory and that the laws of Nature entail no feedback. Physics assumes that all objects in the universe are passive and response to a stimulus does not affect the internal state of the object: objects are non-organizing, the opposite of self-organizing objects. To understand the behavior of a physical object, one needs to analyze its constituents: the object is made of molecules, which are made of atoms, which are made of leptons and quarks, which are made of...

There is no end to this type of investigation, as history has proved. The behavior of matter still eludes physicists even if they have reached a level of detail that is millions of times finer-grained than the level at which we operate. There is no end to this type of investigation, because everything has constituents: there is no such thing as a fundamental constituent. Just like there is no such thing as a fundamental instant of time or point of space. We will always be able to split things apart with more powerful equipment. The equipment itself might be what creates constituents: atoms were "seen" with equipment that was not available before atoms were conceived.

In any case it is the essence itself of a "reductionist" (constituent-oriented) science that requires scientists to keep going down in levels of detail. No single particle, no matter how small, will ever explain its own behavior. One needs to look at its constituents to understand why it behaves the way it behaves. But then it will need to do the same thing for each new constituent. And so on forever. Over the last century, Physics has gotten trapped into this endless loop.

Could matter in general be analyzed in the same way that we analyze neural networks? Could matter be explained in terms of self-organizing systems? Neural networks remember and learn. There is evidence that other objects do so too: a piece of paper, if folded many times, will "remember" that it was folded and will learn to stay folded. Could we represent a piece of paper as a self-organizing system?

Nature exhibits a "hierarchy" of sorts of self-organizing systems, from the atomic level to the biological level, from the cognitive level to the astronomical level. The "output" of one self-organizing system (e.g. the genome) seems to be a new self-organizing system (e.g. the mind). Can all self-organizing systems be deduced from one such system, the "mother" of all self-organizing systems?

We are witnessing a shift in relative dominant roles between Physics and Biology. At first, ideas from physical sciences were applied to Biology, in order to make Biology more "scientific". This led to quantifying and formalizing biological phenomena by introducing discussions on energy, entropy and so forth. Slowly, the debate shifted towards unification of Physics and Biology, rather then unidirectional import of ideas from Physics. Biological phenomena just don't fit in the rigid deterministic model of Physics. Then it became progressively clear that biological phenomena cannot be reduced to Physics the way we know it. And now we are moving steadily towards the idea that Physics has to be changed to cope with biological phenomena, it has to absorb concepts that come from Biology.

In order to accommodate biological concepts, such as selection and feedback, in order to be able to encompass neural and living systems, which evolve in a Darwinian fashion and whose behavior is described by non-linear equations, Physics will need to adopt non-linear equations and possibly an algorithm-oriented (rather than equation-oriented) approach.

Almost all of Physics is built on the idea that the solution to a problem is the shortest proof from the known premises. The use and abuse of logic has determined a way of thinking about nature that tends to draw the simplest conclusions given what is known (and what is not known) about the situation. For example, it was "intuitive" for scientists to think that the immune system creates anti-bodies based on the attacking virus. This is the simplest explanation, and the one that stems from logical thinking: a virus attacks the body, a virus is killed by the body; therefore the body must be able to build a "killer" for that virus. The disciplines of life constantly remind us of a different approach to scientific explanation: instead of solving a mathematical theorem through logic, nature always chooses to let things solve themselves. In a sense, solutions are found by natural systems not via the shortest proof but thanks to redundancy. The immune systems creates all sorts of antibodies. An invading virus will be tricked into "selecting" the one that kills it. There is no processor in the immune system that can analyze the invading virus, determine its chemical structure and build a counter-virus, as a mathematician would "intuitively" guess. The immune system has no ability to "reason" about the attacking virus. It doesn't even know whether some virus is attacking or not. It simply keeps producing antibodies all the time. If a virus attacks the body, the redundancy of antibodies will take care of it.

This represents a fundamental shift of paradigm in thinking about Nature. For many centuries, humans have implicitly assumed that the universe must be behaving like a machine: actions follow logically from situations, the history of the universe is but one gigantic mathematical proof. It is possible that the larger-scale laws of nature resemble very little a mathematical proof. They might have more to do with randomness than with determinism.

The distinction between instruction and selection is fundamental. Physics has evolved around the concept of instruction: mathematical laws instruct matter how to behave. Selection entails a different set of mind: things happen, more or less by accident, and some are "selected" to survive. The universe as it is may be the product of such selection, not of a logical chain of instructions.

Physics is meandering after the unified theory that would explain all forces. What seems more interesting is a unification of physical and biological laws. We are now looking for the ultimate theory of nature from whose principles the behavior of all (animate and inanimate) systems can be explained. Particles, waves and forces seem less and less interesting objects to study. Physics has been built on recurring "themes": planets revolve around the sun, electrons revolve around the nucleus; masses attract each other, charged particles attract each other. Still, Physics has not explained these recurring patterns of Nature. Biology is explaining its recurring patterns of evolution.

A new scenario may be emerging, one in which the world is mostly non-linear. And somehow that implies that the world self-organizes. Self-organizing systems are ones in which very complex structures emerge from very simple rules. Self-organizing systems are about where regularity comes from. And self-organizing systems cannot be explained by simply analyzing their constituents, because the organization prevails: the whole is more than its parts.

## *The Universe as the Messenger*

One pervasive property of the universe and everything that exists is communication. Things communicate all the time.

The Austrian physicist and philosopher Ernst Mach, held in great consideration by Einstein, had a vision of the universe that proved influential on all $20^{th}$ century Physics. Newton defined inertial systems as systems which are not subject to any force. They move at constant or null speed. Systems that are accelerated are not inertial and, by magic, strange forces ("inertial forces") appear in them. Mach realized that all systems are subject to interactions with the rest of the universe and redefined inertial systems as systems which are not accelerated in the frame of fixed stars (basically, in the frame of the rest of the universe). The inertia of a body is due to its interaction with the rest of the matter in the universe.

Mach's principle implies that all things communicate with all other things all the time. This universe appears to be built on messages.

The dynamics of the universe is determined to a large extent by the messages that are exchanged between its parts (whether you look at the level of RNA, synapses or gravitation).

Things communicate. It is just their nature to communicate. More: their interactions determine what happens next. Things communicate in order to happen. Life happens because of communication. We think because of communications.

If all is due to messages, a theory of the universe should de-couple the message from the messenger.

Messages can be studied by defining their "languages". Maybe, just maybe, instead of sciences like Physics and Biology we should focus on the "language" of the universe.

### *Not in Our Name*
If one takes the Copenhagen interpretation of Quantum Mechanics to the letter, in the beginning there were only probabilities.

There was no observer to "collapse" the wave functions of whatever existed, therefore there were only evolving wave functions: probabilities.

There was, technically speaking, no reality to talk about, because there was no observer to cause it to collapse to the state of reality.

Human reality depends on humans to exist and to observe it.

Then at some point, out of this giant network of probabilities a conscious observer was created.

That conscious observer was not a probability: it was a reality.

That conscious observer (the human race) is making the entire universe metamorphose from probabilities into reality.

In this story there seems to be something missing: how do probabilities create a conscious observer?

How does a state of probabilities evolve into a state of conscious observers?

Was the conscious observer merely a probability from the beginning, that turned out to exist out of sheer luck, or was it meant to exist from the beginning?

Is everything that can possibly exist eventually going to exist, given enough time, given enough trials and errors?

Is today's conscious observer still only a probability herself, and not an actual "reality"?

Relativity tells a different story, a story of pure determinism: our future is determined by the past.

If Relativity and Quantum Mechanics are both true (despite the fact that we are not capable of explaining one with the other), then the emergence of the observer must have been written in the original conditions, in the original set of probabilities.

The story of the universe may be the story of an entity that is slowly transforming itself from one kind of substance to another kind of substance, and consciousness might be just the tool that it is employing in order to achieve that transition.

### *The Science of Impossibility: The End of Utopia*
It is intriguing that the three scientific revolutions of the last century all involved introducing limits to classical Physics. Newton thought that signals could travel at infinite velocities, that position and momentum could be measured simultaneously and that energy could be manipulated at

will. Relativity told us that nothing can travel faster than the speed of light. Quantum Mechanics told us that we cannot measure position and momentum simultaneously. Thermodynamics told us that every manipulation of energy implies a loss of order. There are limits in our universe that did not exist in Newton's ideal universe. In particular, Relativity, Quantum Physics and Thermodynamics pose limitations on the speed, amount and quality of information that can be transmitted in a physical process.

These limits are as arbitrary as laws and constants. Why these and not others? Could they be just clues to a more general limit that constrains our universe? Could they be simply illusions, due to the way our universe is evolving?

Newton's world has been shaken to its foundations by Darwin's revolution. Natural systems look different now. Not monolithic artifacts of logic, but flexible and pragmatic side effects of randomness. By coincidence, while Physics kept introducing limits, Biology has been telling us the opposite. Biological systems can do pretty much anything, at random. The environment makes the selection. We have been evangelized to believe that nothing is forbidden in Nature, although a lot will be suppressed.

Once all these views are reconciled, Newton's Utopia may be replaced by a new Utopia, with simple laws and no constraints. But it's likely to look quite different from Newton's.

Where to, Albert?

### Further Reading
Ashtekar, Abhay: CONCEPTUAL PROBLEMS OF QUANTUM GRAVITY (Birkhauser, 1991)
Barbour, Julian: THE END OF TIME (Oxford Univ Press, 2000)
Bohm, David: THE UNDIVIDED UNIVERSE (Routledge, 1993)
**Bohm, David: WHOLENESS AND THE IMPLICATE ORDER (Routledge, 1980)**
Bondi, Hermann: "Negative Mass in General Relativity" (1957)
Bohr, Niels: ATOMIC THEORY AND THE DESCRIPTION OF NATURE (Cambridge University Press, 1934)
Bunge, Mario: QUANTUM THEORY AND REALITY (Springer, 1967)
Carroll, Sean: From Eternity to Here (Dutton, 2010)
**Close, Frank: NOTHING (Oxford Univ Press, 2009)**
**Davies, Paul: ABOUT TIME (Touchstone, 1995)**
Deutsch, David: THE FABRIC OF REALITY (Penguin, 1997)
**Deutsch, David: THE BEGINNING OF INFINITY (Viking, 2011)**
Ferris, Timothy: THE WHOLE SHEBANG (Simon And Schuster, 1997)
Flood, Raymond & Lockwood, Michael: NATURE OF TIME (Basil Blackwell, 1986)

**Gell-Mann, Murray: THE QUARK AND THE JAGUAR (W.H.Freeman, 1994)**
Greene, Brian: THE ELEGANT UNIVERSE (WW Norton, 1999)
Guth, Alan: THE INFLATIONARY UNIVERSE (Helix, 1997)
Hawking, Stephen: A BRIEF HISTORY OF TIME (Bantam, 1988)
Heisenber, Werner: THE REPRESENTATION OF NATURE IN CONTEMPORARY PHYSICS (Deadalus, 1958)
**Heisenberg, Werner: PHYSICS AND PHILOSOPHY (1958)**
Hoyle, Fred, Burbidge Geoffrey, and Narlikar Jayant: A DIFFERENT APPROACH TO COSMOLOGY (Cambridge Univ Press, 2000)
**Kaku, Michio: HYPERSPACE (Oxford University Press, 1994)**
Kastner, Ruth: "The Transactional Interpretation of Quantum Mechanics" (Cambridge Univ Press, 2013)
Kuhlmann, Meinard: THE ULTIMATE CONSTITUENTS OF THE MATERIAL WORLD (Ontos Verlag, 2010)
Linde, Andrei: PARTICLE PHYSICS AND INFLATIONARY COSMOLOGY (Harwood, 1990)
Linde, Andrei: INFLATION AND QUANTUM COSMOLOGY (Academic Press, 1990)
Lloyd, Seth: PROGRAMMING THE UNIVERSE (Knopf, 2006)
**Penrose, Roger: THE EMPEROR'S NEW MIND (Oxford Univ Press, 1989)**
**Price, Huw: TIME'S ARROW AND ARCHIMEDES' POINT (Oxford University Press, 1996)**
Prigogine, Ilya: FROM BEING TO BECOMING (W. H. Freeman, 1980)
Puthoff, Harold: GRAVITY AS A ZERO POINT FLUCTUATION FORCE (1989)
**Randall, Lisa: WARPED PASSAGES - UNRAVELING THE MYSTERIES OF THE UNIVERSE'S HIDDEN DIMENSIONS (2006)**
Rees, Martin: BEFORE THE BEGINNING (Simon And Schuster, 1996)
Rosenblum, Bruce & Kuttner, Fred: QUANTUM ENIGMA (Oxford Univ Press, 2006)
Sachdev, Subir: QUANTUM PHASE TRANSITIONS (Cambridge University Press, 2001)
Sakharov, Andrei: VACUUM QUANTUM FLUCTUATIONS IN CURVED SPACE AND THE THEORY OF GRAVITATION (1968)
Scott, Alwyn: STAIRWAY TO THE MIND (Copernicus, 1995)
Smolin, Lee: THREE ROADS TO QUANTUM GRAVITY (Weidenfeld and Nicolson, 2000)
Susskind, Leonard: THE COSMIC LANDSCAPE (Little & Brown, 2005)
Thorne, Kip: Black Holes and Time Warps (Norton, 1994)
Von Neumann, John: DIE MATHEMATISCHE GRUNDLAGEN DER QUANTENMECHANIK/ MATHEMATICAL FOUNDATIONS OF QUANTUM MECHANICS (Princeton University Press, 1932)

Webb, John: THE CONSTANTS OF NATURE (Pantheon, 2002)
**Weinberg, Steven: DREAMS OF A FINAL THEORY (Pantheon, 1993)**
Wheeler, John: At Home in the Universe (American Institute of Physics, 1994)
Wigner, Eugene: SYMMETRIES AND REFLECTIONS (Indiana Univ Press, 1967)
Wilczek, Frank: THE LIGHTNESS OF BEING (Basic Books, 2008)

# A Timeline of Modern Science and Technology

1850: Rudolf Clausius discovers entropy
1851: Armand Fizeau measures the speed of light
1852: Leon Foucault invents the gyroscope
1852: William Thompson "Kelvin" discovers the second law of Thermodynamics ("On a Universal Tendency in Nature to the Dissipation of Mechanical Energy")
1854: Bernhard Riemann's "Habilition Dissertation" introduces manyfolds in a non-Euclidean geometry
1855: Henry Bessemer invents the Bessemer converter for mass-producing steel
1856: Bessemer converter for mass-producing steel
1856: William Perkin, still a teenager, invents the first synthetic dye, mauve
1857: George Pullman invents the bus
1858: August Moebius discovers the "Mobius Strip"
1858: The first transatlantic cable between the USA and Britain
1859: Charles Darwin publishes "The Origin of Species"
1859: Gaston Plante invents the lead-acid cell, the first rechargeable battery
1863: Dmitri Mendeleev's periodic table of elements (56 known elements and a law for discovering the next ones)
1863: Paul Broca proves that language is localized in the left hemisphere of the brain
1864: James Maxwell discovers the laws of electromagnetism
1865: Louis Pasteur discovers that diseases are caused by germs (birth of modern medicine)
1865: Giovanni Caselli launches the first commercial telefax service (between Paris and Lyon)
1865: Gregor Mendel discover the unit of transmission of traits in living beings (the gene)
1866: The first practical dynamo is developed by Siemens
1866: Ernst Haeckel argues that "ontogeny recapitulates phylogeny"
1867: Georges Leclanche invents the zinc-manganese battery (forerunner of the alkaline battery)
1867: Alfred Nobel invents dynamite
1868: Christopher Latham Sholes invents the first practical typewriter
1869: John Hyatt's celluloid, the first industrial plastic
1873: Camillo Golgi's "On the Structure of the Brain Grey Matter" describes the body of the nerve cell with a single axon and several dendrites
1878: Willard Gibbs' "On the Equilibrium of Heterogeneous Substances" on statistical mechanics and thermodynamics
1873: Christopher Latham Sholes invents the QWERTY keyboard (1873), which Remington begins to mass produce
1874: Wilhelm-Max Wundt founds experimental Psychology
1874: Karl Wernicke discovers that aphasia is caused by damage in the left temporal lobe
1876: Alexander Bell demonstrates the telephone (for the first time humans in distant places can talk to each other)
1876: Ferdinand Braun discovers semiconductors

1876: John Hughlings Jackson discovers that loss of spatial skills is related to damage to the right hemisphere
1876: Ferdinand Braun discovers semiconductors
1877: Thomas Edison invents the phonograph that records sound on a cylinder (for the first time humans can store and play sounds)
1877: Ludwig Boltzmann founds Thermodynamics and explains entropy
1879: James Ritty invents the cash register
1879: Georg Cantor invents Set Theory
1879: Siemens demonstrates the first electric railway
1879: Albert Michelson discovers that the speed of light in a vacuum is always 299,792,458 meters per second
1879: Walther Flemming discovers mitosis, the process by which living cells divide
1881: David Gestetner invents the stencil duplicator, the first office machine to make duplicates of documents
1882: Robert Koch discovers the bacterium of tuberculosis, which at the time was the cause of one in seven deaths
1882: Etienne-Jules Marey's chronophotography
1884: Charles Parsons' steam turbine (that will open the age of cheap electricity and fast sea travel)
1885: Daimler and Maybach invent the motorcycle
1885: Joseph Swan uses synthetic fiber to make fabrics
1885: John Kemp Starley invents the "safety" bicycle (with gearing, chain drive and low wheels)
1886: Karl Benz builds a gasoline-powered car
1887: Emile Berliner invents the gramophone that records sound on a disc
1887: The first book is printed using Ottmar Mergenthaler's linotype (by the New York Tribune)
1888: Kodak introduces the first consumer camera (for the first time humans can reproduce an image without a painter)
1888: John Dunlop invents the pneumatic tire
1886: Josephine Cochrane invents the dishwasher
1886: Ernst Mach formulates the principle that the inertia of an object is due to the interaction of that object with all the rest of the world's matter
1886: Heinrich Hertz discovers that electromagnetic waves can be generated by an electric circuit and detected by another circuit (in particular, waves at frequencies from 300 GHz down to 3 kHz, or "radio" waves)
1887: Heinrich Hertz discovers the photoelectric effect
1888: Nikola Tesla invents the alternating-current motor
1890: Herman Hollerith builds an electrical tabulating device (the first electrical calculator)
1890: London inaugurates the world's first electrical subway system
1890: William James's "Principles of Psychology"
1890: Henri Poincare proves the recurrence theorem
1891: Santiago Ramon y Cajal proves that the nerve cell (the neuron) is the elementary unit of processing in the brain
1892: Hendrik Lorentz discovers that the atom is not elementary but is made of smaller units that are electrical in nature

1892: Dmitri Ivanovsky discovers the first virus, a non-bacterial pathogen
1892: August Weismann's "germ plasma" theory of genetic heredity
1892: Henri Poincare` founds Chaos Theory
1893: Almirian Decker's alternate-current power plant (alternating current makes it easy to transmit electricity over long distances)
1893: Nikola Tesla holds the first public demonstration of radio communication
1895: The Lumiere brothers invent cinema (for the first time humans can replay what has happened in the past)
1895: Wilhelm Roentgen discovers X rays, light rays that are invisible to the human eye
1896: Antoine Becquerel observes the radioactive decay of atomic nuclei (birth of nuclear science)
1897: Karl Ferdinand Braun builds the first oscilloscope and invents the cathode ray tube
1897: Joseph-John Thompson discovers that electricity is due to the flow of tiny negatively charged particles (discovers the electron)
1897: Bayer's aspirin
1898: Valdemar Poulsen invents the telegraphone, a device for recording telephone conversations
1898: Pierre Curie and Marie Curie isolate the radioactive elements polonium and radium (and coin the word "radioactivity")
1899: Valdemar Poulsen invents the magnetic recorder
1900: Max Planck discovers that atoms can emit energy only in discrete amounts or "quanta" and that the energy of light is proportional to the frequency
1900: Ferdinand von Zeppelin builds the first rigid dirigible
1900: Sigmund Freud's "The Interpretation of Dreams"
1901: Guglielmo Marconi conducts the first transatlantic radio transmission (for the first time humans can send sounds to any place on Earth without any wires)
1902: Willis Carrier invents the air conditioner
1902: Clarence McClung discovers the sex chromosomes
1903: Wilbur and Orville Wright fly the first airplane
1903: Konstantin Tsiolkovsky's "The Exploration of Cosmic Space by Means of Reaction Devices"
1903: Valdemar Poulsen invents an arc transmitter for radio broadcasts
1903: William Bayliss and Ernest Starling discover that hormones are chemical messengers
1904: John Fleming uses a diode to detect radio signals
1905: Albert Einstein publishes "The Special Theory of Relativity"
1905: Albert Einstein explains that the photoelectric effect is due to the fact that light is made of packets (later dubbed "photons") that behave like particles and its energy can change only by multiples of Planck's constant proportional to the light's frequency
1905: Albert Einstein explains Brownian motion, and proves the existence of atoms
1905: Alfred Binet and Theodore Simon develop the Intelligence Quotient test
1906: William Bateson names a new discipline, "Genetics"
1906: Robert von Lieben invents the triode, the "vacuum tube" (birth of electronics)

1907: Lee DeForest creates the first electronic amplifier
1907: Hermann Minkowski's four-dimensional spacetime
1907: Leo Baekeland invents "bakelite", the first entirely synthetic plastic
1908: Jacques Brandenberger invents cellophane
1908: Ernst Zermelo founds axiomatic set theory
1911: Heike Kamerlingh Onnes discovers superconductivity
1911: General Electric introduces the first commercial refrigerator
1911: Ernest Rutherford discovers that the atom is made of a nucleus and orbiting electrons, and mostly empty, a miniature solar system
1911: Edward Thorndike founds "connectionism" to explain how the mind learns
1912: Alfred Wegener discovers the continental drift
1912: Joseph John Thomson invents the mass spectrometer
1912: Max Wertheimer founds Gestalt Psychology
1913: Ford installs the first assembly line
1913: John Watson founds Behaviorism
1913: Niels Bohr proves that electrons are permitted to occupy only some orbits around the nucleus of the atom, and the angular momentum of an electron is proportional to Planck's constant, and the energy of an atom changes in discrete quantities
1915: Albert Einstein publishes "The Theory of General Relativity"
1916: Karl Schwarzschild predicts the existence of black holes
1917: Wolfgang Koehler studies problem solving in chimpanzees
1918: Ronald Fisher founds Population Genetics
1918: Hermann Weyl introduces the concept of gauge field to unify gravitation and electromagnetism
1919: Theodor Kaluza adds a fifth dimension to General Relativity
1920: David Hilbert sets out a program to axiomatize mathematics
1921: Edward Sapir formulates the "principle of linguistic relativity" that the structure of a language affects the ways in which its speakers think
1923: Jean Piaget formulates the theory that the mind grows just like the body grows
1923: Arthur Holly Compton performs an experiment (the "Compton Effect") demonstrating that light cannot be only a wave but must also be made of particles
1924: Louis DeBroglie discovers that matter is both particles and waves, with frequency and wavelength being proportional to energy and momentum
1924: Alexander Oparin formulates the theory of the "primordial soup" to explain the beginning of life
1924: Hans Berger records electrical waves from the human brain, the first electroencephalograms
1924: Otto Laporte formulates the law of conservation of parity
1925: Satyendra Nath Bose and Albert Einstein discover a condensate that exhibits macroscopic quantum phenomena
1925: Wolfgang Pauli discovers that some particles (the "fermions") can never occupy the same state at the same time
1926: Erwin Schroedinger's equation of Quantum Mechanics
1926: Oskar Klein proposes a fourth spatial dimension that is undetectable because it is the size of the Planck length
1926: Films with synchronized voice and music are introduced (talking movies)

1926: Robert Goddard launches the first liquid-fuel rocket
1926: Max Born's probabilistic interpretation of the wave amplitudes in Schroedinger's equation.
1927: First vaccines for tuberculosis and tetanus
1927: Philo Farnsworth invents the television
1927: Werner Heisenberg discovers the uncertainty principle
1927: Louis de Broglie discovers a "hidden-variables" interpretation of Quantum Mechanics
1927: Fritz London introduces the first successful gauge theory (phase invariance of electromagnetism)
1928: Paul Dirac discovers antimatter
1928: Fritz Pfleumer invents magnetic tape for audio recording
1928: Alexander Fleming discovers penicillin
1928: Umberto Nobile's dirigible flies over the North Pole
1929: Edwin Hubble discovers that the universe is expanding
1930: Karl Lashley discovers that functions are not localized but distributed around the brain
1930: Paul Dirac proves that the vacuum is not empty
1930: Wolfgang Pauli derives theoretically the existence of the neutrino, a particle that does not interact with ordinary matter
1931: Kurt Goedel's theorem of incompleteness
1932: Fredrick Bartlett formulates the theory of Reconstructive Memory
1932: James Chadwick discovers the neutron
1933: Edwin Armstrong invents FM radio
1933: Ernst Ruska builds an electron microscope that exceeds the resolution attainable with an optical microscope
1935: Wallace Carothers invents nylon
1935: Robert Watson-Watt builds the first RADAR
1935: AEG introduces the first magnetic tape recorder
1935: Albert Einstein, Boris Podolsky and Nathan Rosen discover an apparent paradox of Quantum Mechanics (the EPR paradox)
1935: Arthur George Tansley introduces the concept of the "ecosystem"
1936: Technetium, the first human-made element
1936: Alan Turing's Universal Machine
1936: Heinrich Focke flies the first helicopter
1937: Chester Carlson invents the photocopier
1938: Otto Hahn, Fritz Strassman and Lise Meitner demonstrate nuclear fission
1938: Chester Carlson invents xerography
1939: Niels Bohr and John Wheeler describe the mechanism of nuclear fission
1939: Walter Schottky explains how the interface between a semiconductor and a metal works
1943: Enrico Fermi achieves a nuclear reaction
1943: Tommy Flowers and others build the Colossus, the world's first programmable digital electronic computer
1944: Oswald Avery discovers that genes are made of DNA
1945: Howard Florey and Ernst Chain develop the first antibiotics
1945: John Von Neumann designs a computer that holds its own instructions, the "stored-program architecture"

1945: The first atomic bombs are exploded by the USA
1947: John Bardeen and William Shockley invent the transistor
1947: Edwin Land invents Polaroid, the first instant camera
1947: Norman Wiener's Cybernetics
1947: Dennis Gabor invents the hologram
1948: Claude Shannon's Theory of Information
1948: Hendrik Casimir shows how the zero-point energy can be detected ("Casimir effect")
1948: Georgiy Gamow develops the Big Bang theory
1949: Donald Hebb's neural selectionism
1949: John von Neumann computes pi to 2,037 decimal places using the ENIAC computer
1950: James-Jerome Gibson argues that biological systems pick up information from the environment
1950: "Human calculator" Shakuntala Devi tours Europe
1951: Carl Djerassi and others invent the oral contraceptive pill
1951: William Wilson Morgan discovers the structure of the MilkyWay galaxy
1951: David Bohm hypothesizes that Quantum Mechanics requires a fifth dimension
1951: Electricity is generated by a nuclear reactor at Arco in Idaho
1952: Harold Urey and Stanley Miller recreate the conditions of early Earth in a laboratory and show how aminoacids may have formed
1953: Eugene Aserinsky discovers "rapid eye movement" (REM) sleep that corresponds with periods of dreaming
1953: Francis Crick and James Watson discover the double helix of the DNA
1953: Taiichi Ohno invents "lean manufacturing" (or "just-in-time" manufacturing)
1953: Roger Sperry studies the "split brain" and discovers that the two hemispheres are specialized in different tasks
1954: George Devol designs the first industrial robot, Unimate
1954: Chen Ning Yang and Robert Mills generalize Maxwell's electromagnetism
1954: The Obninsk Nuclear Power Plant in the USSR became the first nuclear power plant to generate electricity for a power grid
1954: Bell Labs' Gerald Pearson, Calvin Fuller and Daryl Chapin build the first silicon solar cell
1954: The first transistor radio ("Regency")
1955: John McCarthy's Artificial intelligence
1955: Jonas Salk develops the first polio vaccine
1955: Niels Jerne proposes a natural-selection theory of antibody formation
1956: Charles Ginsburg builds the first practical videotape recorder
1956: The first flying car, the Aerocar, is certified in the USA
1956: Chien-Shiung Wu, Chen Ning Yang and Tsung-Dao Lee prove the violation of parity
1957: Frank Rosenblatt conceives the "Perceptron", the first artificial neural network
1957: Albert Sabin develops the oral polio vaccine
1957: The Soviet Union tests the R-7 Semyorka, the first intercontinental ballistic missile (ICBM)

1957: John Bardeen, Leon Neil Cooper, and John Robert Schrieffer provide a theory explaining superconductivity
1957: Hugh Everett introduces an interpretation of Quantum Mechanics without uncertainties, the multiverse
1957: Noam Chomsky's theory of grammar
1957: the Soviet Union launches the first artificial satellite, the Sputnik, mostly designed by Sergei Korolev
1958: Boeing introduces the long-distance jet
1958: Jack Kilby invents the integrated circuit
1958: Jim Backus invents the Fortran programming language, the first machine-independent language
1959: Eveready (later renamed Energizer) introduces the alkaline battery
1959: Michel Jouvet discovers that REM sleep is generated in the brain stem
1959: Robert Noyce co-invents the integrated circuit
1959: Min Chueh Chang invents in-vitro fertilization
May 1960: Theodore Maiman demonstrates the first working LASER
1960: Wernher von Braun spearheads development of NASA's Mercury and Apollo space programs
1961: Charles Bachman develops the first database management system
1961: Marshall Nirenberg and Heinrich Matthaei discover how the 4-letter genetic code gets translated into the 20-letter language of proteins
1961: Fernando Corbato builds the first time-sharing system that allows users to remotely access a computer
1961: Yuri Gagarin becomes the first astronaut
1961: Marshall Nirenberg cracks the genetic code
1961: Jacques Monod and Francois Jacob discover gene regulation
1962: Telstar, the first telecommunication satellite
1962: The first Search for Extraterrestrial Intelligence (SETI) takes place with, among others, Frank Drake and Carl Sagan
1963: Murray Gell-Mann's theory of quarks, Quantum Chromodynamics
1963: The touch-tone phone
1963: Douglas Engelbart builds the first "mouse"
1963: Ivan Sutherland demonstrates "Sketchpad", the first program with a graphical user interface
1964: American Airlines' SABRE reservation system is the first online transaction processing
1964: John Young proposes a "selectionist" theory of the brain (learning is the result of the elimination of neural connections)
1964: John Stuart Bell solves the EPR paradox
1964: IBM introduces the first "operating system" for computers
1964: Japan inaugurates the first "bullet train", the Shinkansen
1964: Peter Higgs proves the existence of a mass-giving boson
1965: DEC introduces the first mini-computer based on integrated circuits, the PDP-8
1965: Robert Holley discovers transfer RNA
1965: Arno Penzias and Robert Wilson discover the cosmic microwave background radiation
1966: Hironari Miyazawa proposes a supersymmetry relating mesons and baryons

1966: Rene Thom formulates catastrophe theory
1967: Jack Kilby develops the first hand-held calculator
1967: The first pulsar is observed
1967: Christian Barnard performs the first human heart transplant
1967: Ilya Prigogine shows that biological systems are dissipative systems which self-organize far from equilibrium
1968: Barclays Bank installs networked "automated teller machines" or ATMs
1968: Andries van Dam introduces the "Undo" command
1968: The Arpanet (Internet) is inaugurated
1968: Gabriele Veneziano discovers that a string can describe the interaction of strongly interacting particles
1969: Neil Armstrong is the first human to walk on the Moon
1969: The Concorde, a supersonic passenger airplane
1969: Paul MacLean proposes the theory of the "triune brain"
1969: Yoichiro Nambu introduces string theory
1970: The first practical optical fiber is developed by glass maker Corning Glass Works
1970: Michael Gazzaniga and Joseph Ledoux discover the left-brain "interpreter"
1971: Ananda Chakrabart develops a genetically engineered organism, a new species of Pseudomonas bacteria
1971: Sony introduces the U-matic, first commercial videocassette recorder (VCR)
1971: Ted Hoff and Federico Faggin build the first universal micro-processor
1971: Pierre Ramond introduces the first supersymmetric theory
1972: Ray Tomlinson invents e-mail
1972: Robert Moore and Irving Zucker discover that the suprachiasmatic nuclei is the site of the circadian biologic clock
1972: Hamilton Watch introduces the Hamilton Pulsar P1, the first electronic digital watch and the first using a digital LED display
1972: Raymond Damadian builds the world's first magnetic resonance imaging (MRI) machine
1972: Godfrey Hounsfield and Allan Cormack invent computed tomography scanning or CAT-scanning
1972: Theodore Friedmann and Richard Roblin's "Gene Therapy for Human Genetic Disease?"
1972: Paul Berg's team synthesizes the first recombinant DNA molecule
1972: The Global Positioning System (GPS) is launched
1972: Magnavox introduces the first videogame console, "Odyssey"
1973: Sharp develops the LCD technology for display monitors
1973: Stanley Cohen and Herbert Boyer create the first recombinant DNA organism (the birth of "biotechnology")
1973: Brandon Carter introduces the "anthropic principle" in cosmology
1973: Martin Cooper invents the cellular telephone
1973: Jean-Pierre Changeux discovers neural Darwinism
1974: Ed Roberts invents the first personal computer, the Altair 8800
1974: Sam Hurst invents the touch-screen user interface
1974: Stephen Hawking discovers the radiation of black holes

1974: John Schwarz suggests that string theory is a theory of gravity (superstring theory)
1974: Howard Georgi and Sheldon Glashow propose a grand unification theory (GUT) to unify weak, strong and electromagnetic forces
1975: Benoit Mandelbrot presents a theory of "fractals"
1975: Wilson Edward Osborne founds Sociobiology
1976: Martin Hellman, Ralph Merkle and Whitfield Diffie describe the concept of public-key cryptography
1976: Julian Jaynes introduces the theory of the "bicameral mind"
1976: Sergio Ferrara, Daniel Freedman, and Peter van Nieuwenhuizen introduce the first supersymmetry that included gravity.
1977: The Voyager unmanned probes are launched to explore the solar system and beyond
1977: The World Health Organization (WHO) announces the eradication of smallpox
1977: Frederick Sanger invents a method for rapid DNA sequencing and publishes the first full DNA genome of a living being
1978: Louise Brown is born through Robert Edwards' technique of in-vitro fertilization, the first "test-tube baby"
1980: Douglas Hofstadter publishes "Godel Escher Bach"
1980: Humberto Maturana publishes "Autopoiesis and Cognition"
1980: Ilya Prigogine publishes "From Being to Becoming"
1980: Alan Guth's inflationary model of the universe
1981: Gerd Binnig and Heinrich Rohrer build the scanning tunneling microscope, an instrument for "seeing" the atomic level
1982: Richard Feynman proposes a universal quantum simulator that can simulate any physical object
1982: Andrei Linde's chaotic inflationary multiverse
1983: Kary Banks Mullis develops the polymerase chain reaction for DNA sequencing
1984: Psion introduces the first personal digital assistant
1984: Barry Marshall and Robin Warren show that ulcers are caused by bacteria
1985: David Deutsch's universal quantum computer
1984: Fujio Masuoka invents flash memory
1984: Michael Green and John Schwarz demonstrate that superstring theory can only work in ten dimensions
1986: The Soviet Union launches the permanent space station MIR
1986: Ernst Dickmanns demonstrates the self-driving car "VaMoRs"
1986: Karl Muller and Johannes Bednorz discover the first high-temperature superconductor
1986: Abhay Ashtekar founds quantum loop theory
1987: Applied Biosystems introduces the first fully automated sequencing machine
1989: Magellan Corporation introduces the first hand-held GPS receiver
1989: Christof Koch discovers that at, any given moment, very large number of neurons oscillate in synchrony and one pattern is amplified into a dominant 40 Hz oscillation
1990: The Hubble space telescope is launched

1990: The first Internet search engine, "Archie"
1990: Tim Berners-Lee invents the HyperText Markup Language "HTML" and demonstrates the World-Wide Web
1990: Dycam introduces the world's first digital camera
1990: William French Anderson performs the first procedure of gene therapy
1992: Calgene creates the "Flavr Savr" tomato, the first genetically-engineered food to be sold in stores
1992: The first text (SMS) message is sent from a phone
1993: Gerard 't Hooft develops the holographic theory
1995: The MP3 standard is introduced for digital video
1995: The top quark, the last missing quark, is finally observed at Fermilab
1995: Edward Witten introduces M-Theory
1995: Eric Cornell and Carl Wieman produce the first Bose-Einstein condensate
1995: Ward Cunningham creates WikiWikiWeb, the first "wiki"
1995: Michel Mayor and Didier Queloz discover an exoplanet, "51 Pegasi b"
1996: Nokia introduces the first "smartphone"
1996: Giacomo Rizzolatti discovers that the brain uses "mirror" neurons to represent what others are doing
1997: Ian Wilmut clones the first mammal, the sheep Dolly
1997: The Mars Pathfinder is the first rover robot on Mars
1997: Toyota begins selling a hybrid car, the Prius
1998: Saul Perlmutter, Brian Schmidt and Adam Riess discover that the expansion of the universe is accelerating (dark energy)
1998: The first handheld devices to read ebooks
1998: George Mitchell employs hydraulic fracturing or "fracking" to extract natural gas from the shale rock of Texas' Barnett Shale
1999: The first social networking platform, Friendster, is launched by Jonathan Abrams
1999: John Pendry discovers a way to create metamaterials
2003: The Human Genome Project is completed, having identified all the genes in human DNA
2003: A NASA probe finds that one side of the universe is hotter than the other
2004: Andrei Geim and Konstantin Novosolev, isolate individual graphene planes
2005: Rice is the first cereal crop to be sequenced (by the International Rice Genome Sequencing Project)
2010: Craig Venter and Hamilton Smith reprogram a bacterium's DNA
2010: Autonomous vehicles drive 13,000 km from Italy to China, the first intercontinental trip ever by autonomous vehicles
2012: Markus Covert simulates an entire living organism (Mycoplasma genitalium) in software
2012: Kiyotaka Miura at Kyoto University invents quartz glass memories that can hold data for millions of years
2012: PAL-V builds a flying car
2014: Floyd Romesberg chemically synthesizes two artificial nucleotides and inserts them into a bacteria thus creating a new genetic alphabet

(See also the "Timeline of Neuroscience" in the book of this series subtitled "Brain")

## Alphabetical Index of Names

Aharonov Yakir: 76
Albert David: 76
Ambjorn Jan: 116, 117
Anglin James: 80
Ashby Ross: 15
Ashtekar Abhay: 114
Bak Per: 25, 29
Barbour Julian: 102
Bardeen John: 97
Bardeen William: 85
Barrow John: 126
Bateson Gregory: 26
Bedingham Daniel: 80
Bednorz Johannes: 97
Bekenstein Jacob: 104, 109, 114
Bell John: 62, 63, 80
Belousov Boris: 19
Blokhinzhev Dmitri : 64
Bohm David: 63, 64, 65, 66
Bohr Niels: 51, 54, 60, 65, 71, 73
Bojowald Martin: 116
Boltzmann Ludwig: 37, 38, 39, 99, 100, 103
Bondi Hermann: 57, 100
Born Max: 53
Bose Satyendranath: 27, 84, 97
Brustein Ram: 109
Buchler Justus: 12
Bunge Mario: 13
Burbidge Geoffrey: 111
Burgess Cliff: 107
Carnot Sadi: 7, 36
Carroll Sean: 109, 113
Casimir Hendrik: 57
Caves Carl: 82
Clauser John: 63
Clausius Rudolf: 36, 37
Close Frank: 123
Cohen Jack: 22
Cooper Leon: 97
Cramer John: 67
Darwin Charles: 5, 10, 24, 134
Davies Paul: 69, 100
DeBroglie Louis: 52
Descartes Rene: 11, 35

Deutsch David: 78, 79
DeWitt Bryce: 92
Dirac Paul: 56, 61, 83, 102
Eigen Manfred: 16
Einstein Albert: 27, 41, 43, 45, 46, 47, 48, 49, 50, 51, 52, 54, 55, 57, 58, 59, 61, 62, 63, 64, 67, 70, 81, 91, 93, 94, 97, 101, 102, 104, 106, 107, 108, 109, 113, 114, 115, 116, 117, 118, 120, 123, 125, 126, 127, 128, 132
Everett Hugh: 77
Faraday Michael: 34
Feigenbaum Mitchell: 10
Fermi Enrico: 84
Feynman Richard: 28, 68, 76, 102
Fitzgerald George: 42
Flamm Ludwig: 104
Fourier Joseph: 56
Freedman Wendy: 19
Freund Peter: 94
Fuchs Christopher: 82
Fuller Buckminster: 18
Galilei Galileo  Galilei, 35
Galilei Galileo: 5, 33, 40, 41, 45, 57, 82, 122
Gambini Rodolfo: 114
Gamow George: 110
Gell Murray -Mann, 24, 40, 83
Georgi Howard: 90
Ghirardi Giancarlo: 80
Gisin Nicolas: 61
Glashow Sheldon: 90
Godel Kurt: 104
Gold Thomas: 108
Goudsmit Samuel: 51
Greenberger Daniel: 62
Gross David: 93
Guth Alan: 110
Hadad Merav: 109
Haken Hermann: 15
Hamilton William: 33
Hawking Stephen: 77, 104, 105, 108, 109

Heisenberg Werner: 52, 54, 55, 56, 60, 61, 63, 64, 65, 68, 70, 71, 74, 75, 91, 105, 107, 120
Herbert Nick: 63
Higgs Peter: 85
Hooft Gerard't: 81
Hopfield John: 29
Horava Petr: 95, 117
Horne Michael: 62
Hoyle Fred: 111
Hubble Edwin: 50, 105, 106
Jacobson Ted: 109, 128
Josephson Brian: 28
Jurkiewicz Jerzy: 116
Kaku Michio: 94
Kaluza Theodor: 91
Kamerlingh Heike Onnes, 96
Karolyhazy Frenkel: 73
Kastner Ruth: 69
Kauffman Stuart: 24, 25, 26, 29, 113, 115
Kerr Roy: 105
Kimble Jeff: 61
Klein Oskar: 92
Koestler Arthur: 10, 11
Kuhlmann Meinard: 121
Kuramoto Yoshiki: 28
Lagrange Luigi: 33
Langton Chris: 23, 26
Laporte Otto: 87
Laszlo Ervin: 14
Lee TsungDao: 88
Lemaitre Georges: 105, 110
Linde Andrei: 111, 122
Lockwood Michael: 77
Loll Renate: 116, 117
Lorentz Hendrik: 41
Lorenz Edward: 8, 9
Lyapounov Aleksander: 22
Mach Ernst: 41, 118, 119, 132
Madore Barry: 19
Maldacena Juan: 117, 119
Mandelbrot Benoit: 10
Markopoulou Fotini: 116
Maturana Humberto: 18
Maxwell James: 34, 35, 88
May Robert: 10
Mills Robert: 88

Milne Arthur: 102
Minkowski Hermann: 41
Müller Karl: 97
Monod Jacques: 24
Montonen Claus: 93
Nambu Yoichiro: 92
Narlikar Jayant: 111
Newton Isaac: 5, 8, 32, 33, 35, 38, 40, 41, 45, 47, 48, 50, 51, 52, 53, 54, 57, 58, 59, 63, 65, 70, 77, 82, 83, 91, 99, 101, 109, 117, 118, 119, 120, 127, 132, 133, 134
Olive David: 93
Pasteur Louis: 98
Pattee Howard: 12
Pauli Wolfgang: 84
Pearle Philip: 80
Penrose Roger: 73, 74, 96, 106, 108, 113, 115
Perlmutter Saul: 106
Peskin Charles: 27
Piaget Jean: 29
Pietronero Luciano: 10
Planck Max: 51, 56, 59, 73
Podolsky Boris: 61
Poincare Henri: 7
Poincaré Henri: 121
Poincaré Henri: 39
Polchinski Joseph: 95
Price Huw: 67
Prigogine Ilya: 15, 19, 20, 21, 22, 29, 39, 80, 103
Puthoff Harold: 118
Reppert Steve: 27
Ricci Gregorio: 108
Riemann Bernhardt: 47
Riess Adam: 106
Rosen Nathan: 61, 120
Rovelli Carlo: 114
Rutherford Ernest: 51
Ryu Shinsei: 119
Sachdev Subir: 97
Sakharov Andrei: 118
Salam Abdus: 89
Salthe Stanley: 10, 12
Salvio Alberto: 85
Scaruffi Piero: 1, 2, 124
Schac Ruediger: 82

Schmidt Brian: 106
Schrieffer John: 97
Schroedinger Erwin: 36, 52, 53, 54, 60, 61, 70, 71, 75, 76, 77, 80, 81, 91, 96, 125
Schwarz John: 92
Schwarzschild Karl: 50
Scott Alwyn: 67
Sen Amitaba: 114
Shaw Robert: 9
Smale Stephen: 9
Smith Quentin: 112
Smolin Lee: 112, 114
Soddy Frederick: 51
Sommerfeld Arnold: 126
Stewart Ian: 22
Strogatz Steven: 28
Strominger Andrew: 94, 109
Strumia Alessandro: 85
Susskind Leonard: 96, 104
Swingle Brian: 119
Takayanagi Tadashi: 119
Tegmark Max: 123
Thom Rene: 21, 22
Thompson DArcy: 21
Thompson William: 36
tHooft Gerard: 94, 104, 117
Thorne Kip: 105
Tipler Frank: 105
Tolman Richard: 106
Tumulka Roderich: 80
Uhlenbeck George: 51

Unruh Bill: 113, 128
VanRaamsdonk Mark: 120
Varela Francisco: 18, 26
Verlinde Erik: 109
VonBertalanffy Ludwig: 14
VonFoerster Heinz: 8
VonNeumann John: 12, 70, 71, 72, 116
Webb John: 126
Weinberg Steven: 89
Welsh David: 27
Weyl Hermann: 88, 108
Wheeler John: 50, 56, 60, 68, 73, 75, 102, 104, 106
Wiener Norbert: 27
Wigner Eugene: 72, 75, 87, 123
Wilczek Frank: 120
Wilson Kenneth: 114
Winfree Arthur: 28
Witten Edward: 93, 95
Wootters William: 61
Worrall John: 121
Wright Sewall: 25
Wu ChienShiung: 88
Yang ChenNing: 88
Yorke James: 8
Yukawa Hideki: 84
Zeeman Erich: 21
Zeilinger Anton: 61, 62, 81
Zermelo Ernst: 39
Zurek Wojciech: 79

www.ingramcontent.com/pod-product-compliance
Lightning Source LLC
Chambersburg PA
CBHW051709170526
45167CB00002B/593